Bookkeeping for Builders

by

Michael C. Thomsett

Craftsman Book Company

6058 Corte del Cedro, P.O. Box 6500, Carlsbad, CA 92008

Library of Congress Cataloging in Publication Data

Thomsett, Michael C.
Bookkeeping for builders.

Includes index.
1. Construction industry--Accounting. I. Title.
HF5686.B7T469 1989 657'.869 89-1235
ISBN 0-934041-42-3

CONTENTS

Chapter 1

THE BOOKS YOU KEEP

Every business needs books and records. It's as simple as that. If you're getting paid for work done and those payments aren't wages, by definition you're in business. You're required by law to keep track of income and expenses and maintain ledgers and accounts. Any business that hasn't kept adequate records is at the mercy of the I.R.S. Under federal law, the I.R.S. can estimate income and assess taxes accordingly if you have no records to show actual income. No one wants to be in that position.

Even if the law didn't require it, there are many good reasons for keeping business records. I'll explain some of these reasons later in this chapter. For now, let me put it this way: Most experienced builders would no more operate without a bookkeeping system than start building without a set of plans. Your books help you make business decisions, help lenders and bonding companies understand your strengths and capacity, help you plan for future growth and prepare for emergencies.

Maybe it's a coincidence, but every growing, prosperous construction company I know has a good bookkeeping system. These companies are as good at recordkeeping as they are at building. They know how the company stands financially and can see a problem developing before it becomes a crisis.

I've also seen some contractors and subcontractors who were much better at building than they were at keeping records. Unfortunately, contractors like that usually aren't in business very long. The most extreme example I know of was a contractor who ran his business literally out of a shoebox. He had almost no written records, just invoices and receipts collected in a shoebox. Many bills were paid with cash, including some wages. There was no record of income received. When money was available, he used it to pay personal expenses. Taxes weren't withheld from wages earned by most employees. Obviously, that couldn't go on indefinitely. And it didn't. The unfortunate part was that the man was an excellent finish carpenter. He took real pride in his company's work. But when it came to bookkeeping, he simply didn't have a clue.

There may have been a time when a contractor could run a business year after year without business records. But that time has passed. Today, everyone in business is expected to keep adequate records. And the faster your construction company is growing, the more prosperous your business is becoming, the more important it is to have good records right now. If you expect to break into the major leagues of construction contracting some day, you must master the fundamentals of bookkeeping.

Fortunately, elementary bookkeeping isn't hard to master. Anyone who can add and subtract, follow a few simple rules and genuinely wants to learn to keep books should have no trouble following my explanations. Many people struggle with bookkeeping and get discouraged by what seem to be complex rules. But bookkeeping doesn't have to be difficult. There's nothing mysterious about bookkeeping procedures — once you learn the steps. I'm not going to suggest that you'll become a bookkeeping whiz overnight and learn to love posting debits and credits. But nearly anyone can learn enough of the basics to keep records that are adequate for a small construction contracting or subcontracting company.

The purpose of this manual is to help you master bookkeeping essentials as quickly and as easily as possible. We'll cover the subject one topic at a time and in logical order so you can follow my explanation, even if you have no prior experience with bookkeeping.

When you've finished the last chapter in this book, you should be able to set up a bookkeeping system that's adequate for your office. You'll also be able to modify and improve a system if one exists. Each chapter ends with a series of self-test questions and exercises. Answers are in the appendix at the back of the book. Use these questions to test your understanding of each chapter. Identify the subjects that aren't clear to you and review that part of the chapter.

Whether you're keeping books for a contractor or are a contractor trying to keep your own books, this manual is for you. I hope you have an immediate need for what I'm going to explain. You'll discover that learning bookkeeping skills is easier when you can use what you learn almost immediately. Hopefully, some of what you learn in this chapter and the next will help you solve bookkeeping problems tomorrow, next week, and next month.

- I'm going to explain both the rules you need to know and the reasons for those rules. That helps you remember what rules apply in each case.

- You'll learn the purpose for each journal and ledger and how it fits into the overall procedure.

- I'll show you how to reduce the amount of bookkeeping work you have to do by using shortcuts I've learned over the years.

- Most important, I'll take you step-by-step through the actual work a bookkeeper might handle in a construction office.

If nothing else, this will be the most practical reference you'll ever find on bookkeeping for contractors.

Bookkeeping Defined

Let's begin by making a distinction between bookkeeping and accounting. The bookkeeper's job is limited to several specific roles:

1) Recording transactions in journals and ledgers.

2) Establishing an "audit trail" — showing where each number came from.

3) Providing up-to-date information about cash, income, expenses and profits.

4) Gathering information needed to prepare preliminary or rough financial statements. These show the asset value of a business and the profit earned on work done.

You can see that the term *bookkeeper* describes quite accurately what bookkeepers do. They keep books. Accounting is different. Accountants interpret the financial information bookkeepers compile. An accountant will study a financial statement and try to anticipate the future, project tax liabilities, identify potential cash flow problems, or advise the owner on how to finance the purchase of equipment.

In some firms, the bookkeeping and accounting roles are woven together, so the line between the two is hard to distinguish. But our emphasis in this book will be bookkeeping— the recording of financial information. The interpretation of that information is a separate task which I covered in another book, *Builder's Guide to Accounting*. An order form for that book appears in the back of this volume.

Bookkeeping isn't the same as accounting. But that doesn't make bookkeepers any less essential in a construction office. The bookkeeper reports how much cash is in the bank, how much is being spent on expenses, and how much money the company has made or lost. That's important information. And when it's both accurate and timely, it's exactly the information every construction contractor or subcontractor needs.

Your Books Measure Your Success

Business success is measured in dollars and cents. If a business is successful, someone is doing something right. You want to do more of it. If a business is losing money, someone has to stop it and start doing something else. Otherwise the business won't survive. No one is sure how successful a business is or isn't until a bookkeeper totals up the figures.

Here are some common indicators of business success. Notice that the figures needed to measure success are all based on totals compiled by bookkeepers.

Return on investment

How much are you getting back on the money invested in your business? Suppose you begin a business with $50,000 in cash. A year later the value of your business has grown to $60,000. That's a $10,000 return on your $50,000 investment. Your gain is 20 percent (10,000 divided by 50,000).

Profit as a percentage of sales

The most common measure of business success is profit as a percentage of sales. If your profit (income minus all costs and expenses) is $10,000 on gross sales of $200,000, the profit is 5 percent (10,000 divided by 200,000).

Against a standard

If your goal is an 8 percent profit every year, how close did you come to your goal? You can also compare results with competitors, or averages for others in your type of work.

Against a plan

The best type of business planning sets monthly goals for income, expense and profits. With good business records, you can compare results in any category with what you expected or planned to achieve.

With previous years

Progress is important in a business. If you made a 3 percent profit last year, your measure of success might be to raise that to 5 percent this year. There's no way to be sure this year is any better than last if you don't have accurate records that make comparisons possible.

Bookkeeping and Profits

No measure of the financial success of your business will be accurate unless the numbers you're using are accurate and the procedures consistent. Here's an example. Suppose you're comparing profits this year or this month with profits for last year or last month. The comparison won't be accurate unless you're comparing apples to apples and oranges to oranges. If you changed the way income or expenses were recorded at the end of the last year, don't try to compare this year's with last year's figures. The same is true when you're comparing results at your company with results at another company.

Unless both companies use the same procedures, comparisons are worthless.

That's why consistency is so important. To get maximum value from your business records, you have to be consistent. You have to follow the procedures used by bookkeepers and accountants throughout the construction industry. Treat income, costs and expenses the same way every year. Follow the rules I'm going to explain and you'll get maximum value for your effort. Anything else may be a waste of time.

Most one- or two-person construction companies need only the most basic bookkeeping system. But as your company grows, the number of monthly transactions will increase. You'll write more checks each month and there'll be more income to record. Your bookkeeping system has to be flexible enough to keep up with company growth.

In the first two months of one builder's operation, he had one job and wrote only seven checks. He had no employees, he received full payment in two installments, and it was easy to identify profits. In fact, he kept the "books" for this period on one page.

Three years later he was working five jobs at the same time and had four crews and one office employee. Most of his jobs were done on credit. He wrote about 75 checks every month, and had open accounts with eight different supply houses. Company growth required better records and more recordkeeping.

More transactions make it harder to keep track of profit. If you have several jobs going at once, you need a set of records for each. Otherwise you can't tell which jobs are profitable and which aren't. And records from all jobs have to be consolidated each month to get company totals. Salaries, rent, and other overhead costs apply to a number of jobs. Without good records, it's hard to know if the company made or lost money, let alone how much.

The more complex your business becomes, the more information you'll need from your bookkeeping system.

- You'll want to identify income and costs for *each* job.

- You must be able to keep track of many transactions.

- You need to set up specialized records for particular types of transactions. For example, keeping track of money owed to you requires one form of record, while summarizing checks you write each month requires another.

- Records have to be up-to-date and accurate, so current profit information is always available.

Types of Bookkeeping Records

There are several levels of information in a bookkeeping system. Journals show detailed information about every transaction: each check written and each bill sent out. These are recorded in specialized journals. The journals are summarized into the general ledger. And finally, a financial statement condenses the information even further.

Journals provide detailed information about every expense. This is important. For tax purposes, a business can deduct only legitimate business expenses.

The full bookkeeping system includes source documents, books of original entry, and books of final entry. Let's look at the meaning of each of these.

Source Documents

Invoices or receipts establish that money is due, or was spent. When you order lumber for a job, the supplier sends you an invoice, showing exactly what you ordered, when it was delivered, and the price. This is a source document for the materials.

Suppose you go to the post office and pay cash for a roll of stamps. You're given a receipt, which serves as the source document for cash paid out. You'd file the receipt in your "paid bills" file when you take money from petty cash or write a company check to reimburse yourself for the cost of the stamps.

Every check written for a business expense must have a source document that verifies the legitimacy of the expense. Sometimes the source document is a lease agreement that covers many checks during the year, rather than an invoice for a single check.

Occasionally, a source document isn't needed. For example, there would be no receipt for rent paid on your office if it were leased under an oral agreement. In this case, the canceled check is your only receipt. In case of an audit, you couldn't produce an invoice, contract or other written proof. But it should be easy to show that the checks were for rent payments.

File your source documents alphabetically by company name (payee). Start a new file each year. If you need to look up a bill from a particular supplier, it's easy to find it in the paid invoices file.

An alphabetic file of invoices paid is essential. But it isn't enough. You'll also need a list of bills paid by category. For instance, suppose you need a summary of material expenses on the Jones job. It might take you hours to find all this information in an alphabetic file. That's why you need a second level of books.

Books of Original Entry

You'll see, as we proceed, that it would be impractical to try to report or measure profits from a huge list of source documents. You need a way to collect information in convenient categories. We'll start with the broadest categories and create a book of original entry for each.

There are three types of books of original entry; the *receipts journal*, the *disbursements journal*, and the *general journal*. We'll discuss each one in detail in later chapters. For now, here's a brief description:

1) The receipts journal is used to record all income and cash receipts. If you grant credit to customers, income isn't received in cash at the time it's earned, but it's still recorded in this journal.

2) The disbursements journal is used for recording all payments. Every check you write is listed in this journal and is assigned an account number to show which account the expense is charged against.

3) The general journal is used to record transactions that don't belong in either the receipts *or* disbursements journal. These include entries to correct errors, to record non-cash expenses, or to record accruals (more on this in Chapter 5).

Some builders will need a subsidiary journal if they have a heavy volume of transactions of a particular type. Suppose you do a lot of residential repair work and bill customers at month-end. You'd have to keep detailed records of accounts receivable.

Your accounts receivable journal will show the date each job is completed and the amount due. Because bills have to be prepared for each customer, you'll want to keep the accounts receivable information in alphabetical order by customer name. This isn't practical on the sequential receipts journal. So you set up a subsidiary ledger for receivables. The total of all entries on this ledger equals what is reported in the more summarized receipts journal. This way, a lot of detail is broken out separately where you can use it for billing. The receipts journal itself is not cluttered.

Books of Final Entry - the General Ledger

Your general ledger shows the least detail. It probably has one page for each account. There are accounts for all assets (properties owned by the business), liabilities (debts), and net worth (the owner's equity in the business). Income, cost and expense accounts are part of the general ledger also.

You post the general ledger (write figures on the appropriate page) from the receipts, disbursements and general journals. Most general ledger accounts will have no more than three entries each month.

A general ledger account will have a balance forward (the balance of the account from the previous month), plus or minus the current month's entries, and an ending balance. For example:

Postage				
Date	**Description**	**Debit**	**Credit**	**Balance**
	Balance forward			312.16
May 12		25.00		337.16

In this example, the balance forward is $312.16. For May, one entry for $25.00 was added to the account. And the ending balance — the new balance forward— is $337.16.

Every account in the general ledger is set up in the same way. In a double-entry bookkeeping system, every account has two sides, a debit side and a credit side. I'll explain this in detail in Chapter 2.

The general ledger is used to prepare financial statements. That's why you'll want to keep it as detail-free as possible. If too much detail gets into the general ledger, it's hard to post and manage. If you have too many transactions in any general ledger account, consider setting up a subsidiary ledger for that account. Then post only totals from the subsidiary account to the general ledger. Chapter 11 shows how to set up and keep subsidiary records.

Financial Statements

Statements aren't actually part of your books. They're drawn from the general ledger and are the result of your bookkeeping efforts. Their usefulness depends on the accuracy of your bookkeeping ledgers and journals.

Since this chapter is intended only as an overview, I'll just sketch the three financial statements. Chapter 8 explains financial statements in detail.

1) Balance Sheet

This is a summary of all assets, liabilities and net worth. Assets are anything that has value. Examples are cash, equipment, and machinery used in your business. Liabilities are money your company owes. And net worth is what's left after subtracting liabilities from assets. This is also known as your *equity* in the business.

2) Income Statement

Also called the *Profit and Loss Statement*, this reports the income, costs, expenses and net profit for a specified period.

3) Cash Flow Statement

Also known as the *Summary of Sources and Application of Funds*, this shows where cash came from (profits, capital invested, sale of assets, or loan proceeds, for example), where it was spent (to buy equipment or reduce a loan balance, for example), and how much your working capital increased or decreased during the year.

You always use the same closing date for financial statements that are prepared together. If your balance sheet reports the assets, liabilities and net worth as of December 31, your income statement and cash flow report would cover the period ending December 31.

The flow of information from source documents through financial statements is shown in Figure 1-1.

A Sample Transaction

We'll examine each of the specialized journals and ledgers and explain them in detail in later chapters. For now, let's trace one transaction through the books.

On September 5, you order lumber that costs $5,000 for use on a job. The supplier will bill you, and you'll pay the bill on October 10.

Source Document

You'll receive an invoice with the delivery. It shows what was delivered and when, and identifies the supplier. Your supervisor verifies that everything on the invoice was received by initialing the document. This invoice is then filed alphabetically by supplier name in the accounts payable file. If an invoice or statement for this delivery is received through the mail, it should

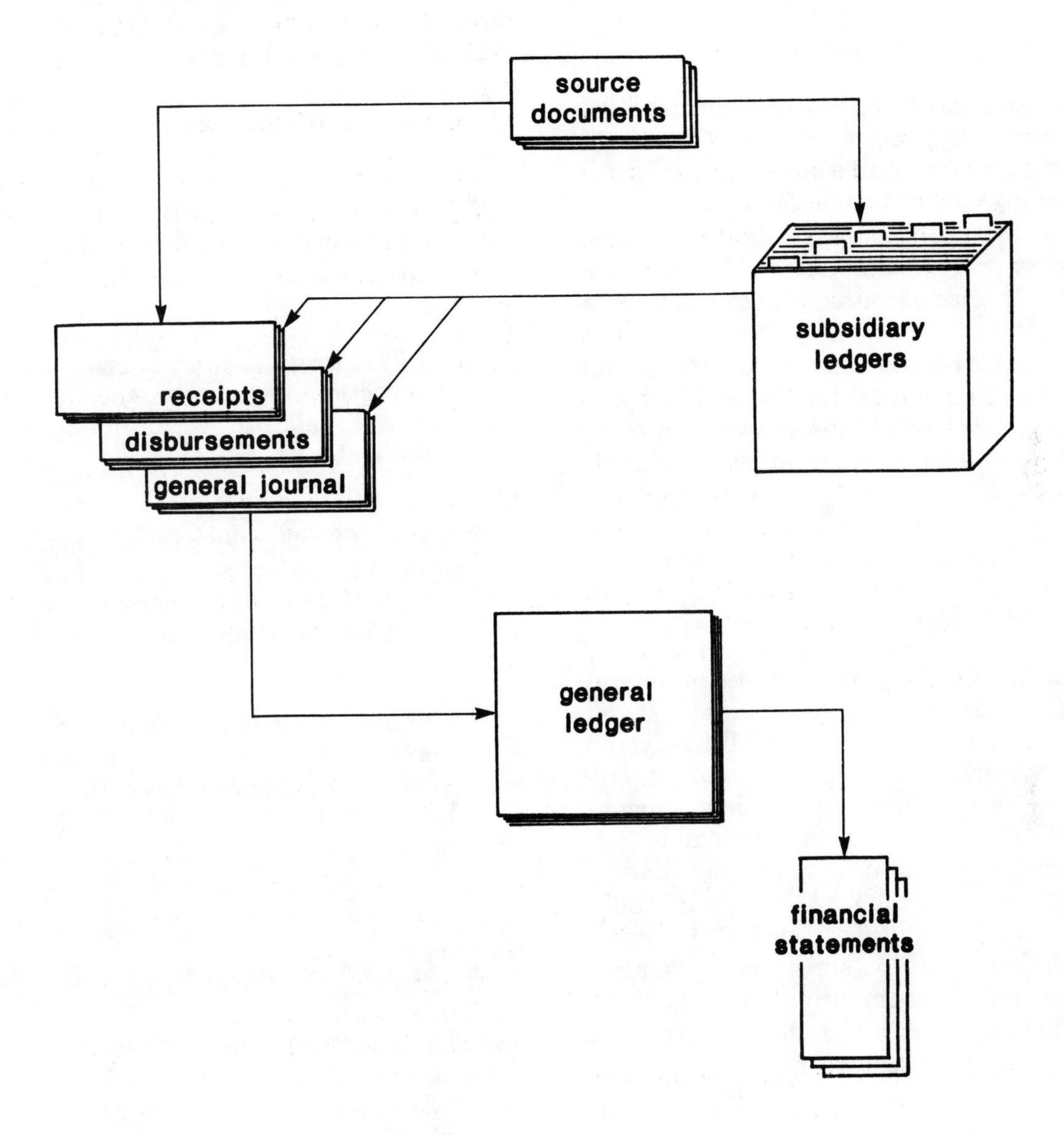
the flow of information
source documents
subsidiary ledgers
receipts
disbursements
general journal
general ledger
financial statements

Figure 1-1

be clipped to the original invoice. When you pay the bill, pull it out of the payables file and refile it (with a copy of your check) in the paid invoices file.

Books of Original Entry

You'd record the payment in the disbursements journal when you pay the bill. You might also enter the payment in a subsidiary ledger if you're keeping one for this account.

You want to record purchases in a purchases journal when you place an order. This is a subsidiary record you'd use to keep track of amounts owed suppliers.

You might also need to record the purchase on the cost record for an individual job. This way you can compare costs to the original estimate and budget, and prepare progress billings to your customer.

Books of Final Entry

You'll summarize the journal entries at the end of a specific period, usually once a month. If you keep a running account of your liabilities, you'd record the material purchase as an account payable (money owed). This way, the debt appears in the books during the month it occurred.

If you don't keep a running tab on your liabilities, you'll record the bill for the materials only when it's actually paid, on October 10. Many firms track their debts in a subsidiary ledger but don't actually enter them into the books except at the end of the year. This reduces paperwork but still allows for accurate reporting for tax purposes.

Financial Statements

Your balance sheet will show the account payable as a debt (liability.) Your income statement will show a purchase of materials. If those materials were bought for inventory rather than for a specific job, there would be no entry on the income statement. Instead, you'd list the debt in the liability section of the balance sheet, and the inventory value in the asset section.

So far, so good. Following one transaction is easy. Complications will accumulate as the number of transactions increases. But if you understand the procedure so far, you're well on the way toward understanding more complex transactions. The procedure is the same. Only the volume is different.

Here's what you should have learned so far in this chapter:

- Source documents prove the validity of a transaction, and are filed alphabetically by payee.

- Books of original entry show details of all transactions of one type, such as cash paid out, liabilities accumulated for the month, etc.

- Books of final entry summarize the transactions in each account, and classify them as increases or decreases to assets, liabilities, income, costs or expenses.

- Financial statements are drawn to show the status in assets, liabilities and net worth. They report your profit (or loss) and show how your cash is being managed.

The Reasons for Keeping Books

You have to record transactions — for income, costs, and expenses — to prepare financial statements. But there are many other advantages to a complete and accurate bookkeeping system. Figure 1-2 illustrates the following six:

Management Information

Anyone who owns or manages a construction company needs to know how the company is doing financially. Some companies need this information only monthly. Others can't always

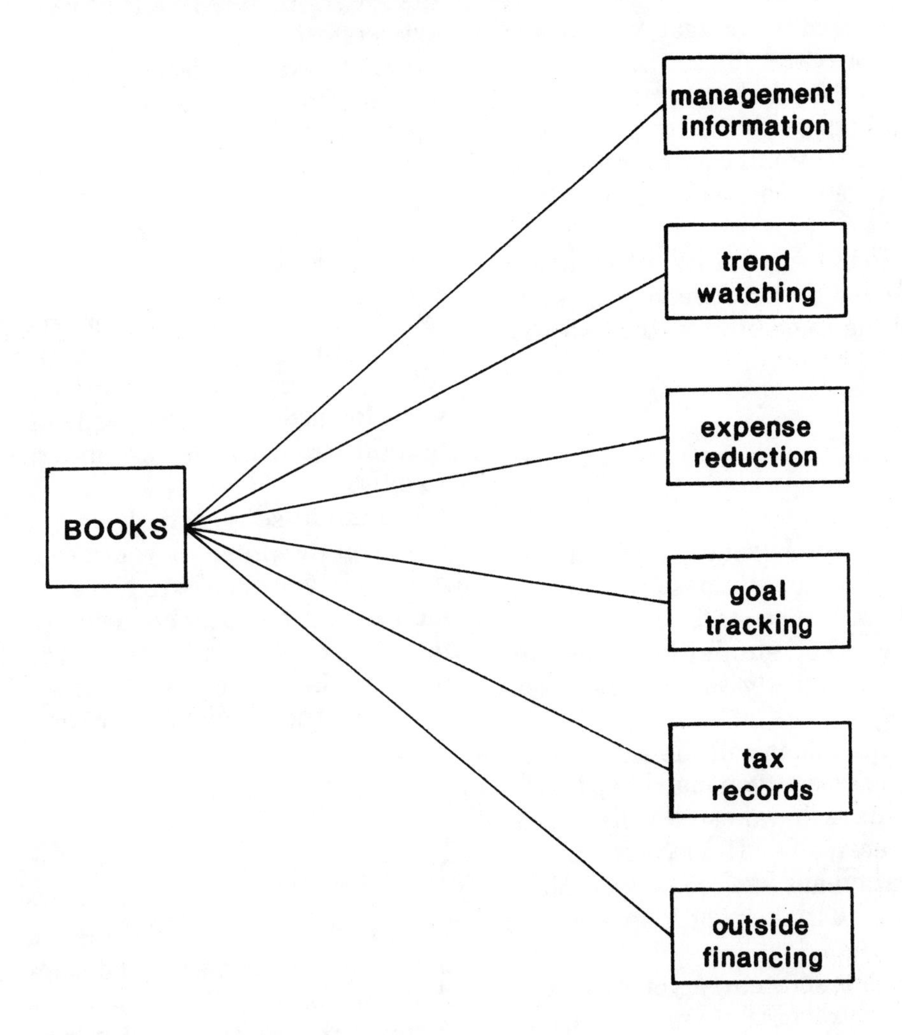
the value of bookkeeping
BOOKS
management information
trend watching
expense reduction
goal tracking
tax records
outside financing

Figure 1-2

wait until the end of the month to review a financial statement. They may need current information weekly, even daily.

Job cost records must be kept up to date. If expenses begin to exceed the budget, you'll have to watch that job more carefully so the company won't lose money on the job. Profit margins can be thin in the construction contracting and subcontracting business. A small cost overrun, idle time at the site, or loss of material, could wipe out the entire profit.

The books have to show accurate information about income and cash flow. You have to be sure customers are paying their bills on time so you can pay yours.

Trend Watching

Results in business are measured by comparisons. How profitable are operations today, compared to last month or last year? Is this job being run as efficiently as other, similar jobs? Are we collecting money as quickly as we did last month?

Answering these questions will show trends in your business. For example, you may keep track of the number of days it takes to collect the average account receivable. If the number of days increases, that means you're not collecting as efficiently as you did in the past. The trend is negative.

If you spot negative trends early, you can usually do something to reverse them before too much damage is done. For example, you could stop extending credit to delinquent customers and demand that accounts be brought up-to-date. The point is, you can control a trend *only after you've found it.*

Expense Reduction

Expenses can easily go over budget if they're not watched closely. One contractor I know suspected he was spending too much on office supplies, auto and telephone expenses. He started a requisitioning system for supplies, required a log of business travel for company-owned vehicles, and changed to a telephone company that gave an account breakdown on each monthly bill. These simple steps saved nearly $1,000 a month — increasing annual profit by 20 percent in the first year. Results like that were well worth the extra effort.

Your books will help you identify problem expense accounts — and make corrections when they're needed.

Goal Tracking

Successful business owners set goals for themselves. And they reach them. But motivation to reach goals comes from comparisons. When you work toward goals, you need summarized information (from the books and records) at least weekly.

You might set yearly goals for income volume based on the previous year. You'd use the information in your books to track the earnings each month and the number of new jobs won each month. Then you'd compare them to the same items for last year. Your goal would be to match or exceed the previous year consistently in both areas.

Tax Records

You must justify everything you claim on your tax return. A good set of books will help you do that. But beyond that, you can sometimes also reduce tax liabilities by timing certain income and expenses. It's completely legal to defer income to the next year, or to prepay many types of expenses to reduce this year's profit.

In some cases, you might be better off reporting as much taxable income as you can in the current year. You might base that decision on changing tax rates or information you have about a large contract coming up next year. You can only make decisions like that if your books are accurate and dependable.

Outside Financing

Any growing business eventually needs to borrow money. As a rule, as long as return on investment is greater than the interest expense, going into debt is a smart business decision.

To qualify for a bank loan, you'll have to provide an accurate financial statement. Your loan officer may want to see the results of operations for two or three years, perhaps more. Your bookkeeping system must be accurate enough to draw up financial statements showing the true picture of business as of the end of any month.

Keep Your Perspective

One caution about the bookkeeping process. As important as a good system is to any contracting business, don't let it become an end in itself. Your books are the means for monitoring business — or taking its temperature and judging how well things are going.

Some bookkeepers develop a narrow point of view and begin to think of the books as a top priority for the company. That's a mistake. Contractors are in the business of contracting. The wise keeper of the books sees the job in perspective within that larger picture. Bookkeeping is a support function. It helps everyone in the company do a better job. It isn't the primary purpose for the business.

If you do the books for an owner, you have to know what really motivates him or her. The owner is concerned with scheduling crews, staying within a project budget, and winning the next contract. A smart owner will use the information you supply to make the company more profitable. But the books aren't the reason for being in business. They're just the foundation that holds up the rest of the house.

If you're a builder and do your own books, you can fall victim to the same pitfall. Don't become obsessed with the bookkeeping. You might create the most efficient set of books possible, but you may end up taking time away from profitable activities like supervising field crews.

There is a middle ground. Recognize the books for their real and important value. But don't lose sight of where the profits are. A good bookkeeping system is fast, up to date, consistent, accurate, and easy. If information is delayed because it takes several days to gather, there's a problem. You need a simple, efficient process that gives you everything you need with the least effort possible.

In the next chapter we'll explain different bookkeeping methods and describe the double-entry bookkeeping system in detail. But don't go on to the next chapter until you can answer the following questions.

Self Test

1. A builder starts a construction business with $80,000 in cash and other assets. A year later, the company income statement shows that sales were $66,000 and net profit was $5,500. The return on investment was:

a) 6.9%

b) 14.5%

c) 8.3%

d) none of the above

2. *Using the same facts as in question 1, the profit on sales was:*

a) 6.9%

b) 14.5%

c) 8.3%

d) none of the above

3. *A check for office supplies would be posted in the:*

a) general journal

b) disbursements journal

c) general ledger

d) all of the above

4. *The total of all entries for receipts would appear in the:*

a) general ledger

b) accounts receivable ledger

c) receipts journal

d) all of the above

5. *A balance sheet shows:*

a) proof of balancing in all accounts

b) assets, liabilities and net worth

c) balance of profits after costs and expenses

d) all of the above

6. *A contractor should refer to the books for:*

a) monitoring of his own goals

b) keeping expenses in line

c) planning ahead for taxes

d) all of the above

Chapter 2

BOOKKEEPING METHODS

There are two accounting systems recognized in modern accounting practice: accrual accounting and cash basis accounting. In this chapter I'll explain those terms and a few others that bookkeepers need to know. I'll also introduce you to the type of work bookkeepers handle in a construction office.

Most construction companies use accrual accounting. Some use cash basis accounting. *The system you follow isn't important.* Here's what is important: Your books must reflect reality. They have to give a true picture of the financial condition of your business. You don't keep books just because the law requires it. You keep financial records to know how your business is doing, to determine what's right and what's wrong about your business, and to provide the information needed to make sound business decisions. That's why every company needs accurate financial records.

Your accountant will help you decide which accounting method is best for your business. Tax law will probably influence the decision. You'll have to choose the accrual system if you maintain inventory, or if your annual gross sales exceed a certain amount. Your accountant will know the choices and will understand the advantages of each system.

Cash vs. Accrual Accounting

Cash accounting is the simplest. You record (or *post*) a transaction only when money changes hands. You don't record income when it's earned, only when the customer actually pays. And regardless of when you buy something, there's no entry on the books until you pay for it.

There's a disadvantage to cash accounting. Any financial statements drawn up based on the cash method will be inaccurate. Consider these situations:

- During one month you finished $54,000 worth of business and billed your customers. But at the end of the month, customers had not yet paid their bills. Because no cash has been received by you, your financial statement doesn't show this income.

- At the end of the month, you owe bills for materials and expenses totaling $28,750. The financial statement reports a profit of $18,000 for the month. But in fact, the operation lost more than $10,000. The statement does not show unpaid bills.

Cash accounting is a poor choice if you aren't paid promptly for work done or if you pay material suppliers and subcontractors well after the money is earned. In some businesses, cash accounting just doesn't reflect reality.

Your accountant might determine that *paying taxes* on the cash basis is best for your company. Even so, you can keep your books on the more accurate accrual method, then adjust the year-end report for tax purposes. You're allowed to keep books and draw financial statements on one basis, and report taxes on another.

When you keep books on an accrual system, you record everything — income, costs and expenses – when it's earned or incurred. It doesn't matter when you actually put the money in the bank, or spend it.

Suppose you finish a job during May, but you don't receive payment until June. With accrual accounting, income is *recognized* (put onto the books) in May, the month it was earned. It's set up as an account receivable. In June, when the payment comes in, accounts receivable is reduced and your cash account is increased.

The same rule applies to costs and expenses. For example, you purchase $1,200 in materials and use them on a job in August. The bill is sent to you at the end of the month. You pay it on September 10. In this case, the cost was incurred in August, and is shown on the books as an account payable, an accrued cost. When you pay the bill in September, you reduce accounts payable.

In both of these cases, financial statements will accurately show profit for the month, even when payments are made and received in later months.

Accrual accounting also works another way. Suppose you have a contract for an $80,000 job. At the end of June, you've completed 60 percent of the project. Under the accrual method, you should report $48,000 in income (60 percent of the total contract). But you've received $60,000 in payments. The excess of $12,000 is called *unearned* income. An accrual would defer (put off) recognition of this income until a future month, when it will become earned. Your financial statement won't show income that should apply to another period.

Another example: To get a discount, you pay three months' rent in advance on a backhoe. The actual expense should be spread over three months. One-third is recognized in the first month, and the balance is set up as a prepaid expense. You reverse the accrual over the next two months.

Accrual entries are made in the general journal, which we'll discuss in detail in Chapter 5. An accrual is a temporary entry. You must eventually reverse it. Every increase to accounts receivable or payable this month must be reduced in the future, when those items are received or paid.

To summarize, there are two types of accrued income and expenses. Accruals which precede the spending of cash are:

- Earned income (income is earned but not received until later).
- Incurred costs or expenses (materials are ordered and received, but not paid for until later).

Examples of accruals following the spending of cash are:

- Unearned income (cash is received now, but not earned until later).
- Prepaid expenses (cash is paid now for costs or expenses not incurred until later).

Accrual accounting adjusts each month's cash transactions to give the most accurate picture of actual profit. Because business is conducted largely on credit, accrual accounting is essential for realistic financial reports.

the single-entry method

DATE	ENTRY	PLUS	MINUS
1-10	income received, Smith	1,000.00	
1-11	materials purchased, check 414		516.32
1-16	telephone bill, check 415		107.40

Figure 2-1

Single-Entry Accounting

One way to keep books would be to write every entry in a journal, either as a plus or a minus. This is called single-entry accounting. You increase the balance by receipts (income) and reduce it for payments (expenses). From the collective entries for one month, you could draw a cash-basis financial statement. If the company made money, the total would be positive. The total would be negative if the company lost money in the month.

An example of single-entry accounting is shown in Figure 2-1. In this simplified example, the builder received one payment and made two payments. The income is a plus (increase to cash) and the two payments are minus (reductions of cash). As long as you have a limited number of transactions, single-entry accounting is an adequate system. But it doesn't reflect work completed or billed out and doesn't show how much is owed. So your financial statements wouldn't give a true picture of what happened during the month.

Another problem with this method is that it's very hard to guarantee mathematical accuracy. Any error would likely go undetected. Because you're only repeating what's already in your business account checkbook, single-entry accounting isn't very useful.

Of course, it's better to keep single-entry records than no records at all. If you're just starting out, you probably don't have a bookkeeper. Chances are you're using your personal checking account and cash for expenses. I don't recommend doing that. You probably haven't taken the time to organize any bookkeeping system at all. But you still need to keep track of income and expenses. The law requires it.

In this case, single-entry accounting is acceptable. Back up each entry with a receipt, invoice or voucher of one kind or another. That way you'll have source documents and a temporary book of original entry. Once you or your accountant set up the permanent books, you can transfer the information from the single-entry journal into a more complete and professional system.

Double-Entry Accounting

In double-entry accounting, as the name implies, you make two offsetting entries for every transaction. If the total of the two sets of entries is the same, the books are said to be *in balance.* In double-entry bookkeeping, no minus signs are needed. Every entry in the journal is a positive number. When there's a posting error, you'll know because the books won't be in balance — the debits don't equal the credits.

In a single-entry system, it's possible to add an account incorrectly and not find the error. The only way to detect it is to double-check all the math in the books.

Every account category in a double-entry system has two sides, a left side or "debit" and a right side or "credit." The posted debits must always equal the posted credits. A math error or exclusion will show up because a properly posted set of books will always add up to zero.

Let's see how the three entries in Figure 2-1 would look in a double-entry system. Each entry will be posted to the debit side of one account, and the credit side of another. Don't worry just now about how to decide which side of each account gets the entry. I'll explain that a little later.

- Income is a debit to the cash account, offset by a credit to sales (income), which is reported as a credit balance. You *increase* income every time you make a credit entry. So a $1,000 credit to income last month, plus a $2,000 credit this month, will add up to $3,000 total income to date.

- Payment for materials is a debit to the direct cost, materials. And the same entry is a credit to the cash balance.

- Paying a telephone bill is a debit to the telephone expense account, and a credit to the cash balance.

Figure 2-2 shows the three entries you'd make on a double-entry system.

Now we'll see how the individual accounts would look with these three entries posted. Note that any number less than zero is put in parentheses. This is an accounting custom.

Cash			
Date	Debit	Credit	Balance
1-10	1,000.00		1,000.00
1-11		516.32	483.68
1-16		107.40	376.28

Income			
Date	Debit	Credit	Balance
1-10		1,000.00	(1,000.00)

Materials			
Date	Debit	Credit	Balance
1-11	516.32		516.32

Telephone Expense			
Date	Debit	Credit	Balance
1-16	107.40		107.40

Now, to prove that all posting is in balance, i.e., that it adds up to zero, simply add up the ending balances:

the double-entry system

DATE	ACCOUNT	DEBIT	CREDIT
	1-1		
1-10	cash	1,000.00	
	revenue		1,000.00
	1-2		
1-11	Materials	516.32	
	cash		516.32
	1-3		
1-16	telephone	107.40	
	cash		107.40

Figure 2-2

Cash	376.28
Income	(1,000.00)
Materials	516.32
Telephone	107.40
Balance	0.00

This proves that amounts posted to all accounts were correct and that the addition or subtraction within accounts is accurate.

This is a simplified example involving only three transactions. But the same principle applies even when you make hundreds of entries every month. With that many possible errors every month, we need a system that helps us spot errors. That's where double-entry bookkeeping really shines.

Identifying Debit and Credit Entries

Each type of account has a normal debit or credit value. For example, cash and other assets are usually debit balance accounts, while liabilities carry a credit balance. Income is always a credit balance, while costs and expenses are debits. Your balance sheet will list most of these accounts in the order shown.

Category	Account	Debit	Credit
Assets	cash	x	
	accounts receivable	x	
	inventory	x	
	fixed assets	x	
	prepaid expenses	x	
	reserve for depreciation		x
Liabilities	accounts payable		x
	notes payable		x
	deferred income		x
Net worth	owner's equity		x
Income	income		x
Costs	materials	x	
	direct labor	x	
Expenses	all expense accounts	x	

Accounts that normally show debit balances will increase on the debit side. Accounts that normally show credit balances will increase on the credit side.

Here's an example to help you reason an entry through. When you receive cash due from a customer, you have to increase two accounts, cash and income. Cash increases on the debit side so you would debit cash. Income increases on the credit side, so you would credit income. When you pay a bill, reduce the cash account by putting the entry on the credit side. The expense account would increase, so its entry would be on the debit side.

You use double entry accounting to accrue non-cash entries. Here are some sample entries in double-entry books.

Event	Debit	Credit
A job is completed and income is earned	accounts receivable	income
The amount due is paid the following month	cash	accounts receivable

The second entry reverses the accrual, so the overall effect on accounts receivable is zero. The account is debited when income is earned, and it's credited when income is received. The accrual entry has increased (credited) income in the month earned. It has also increased (debited) cash in the month the money was received. The same sequence applies for expenses.

Date	Event	Debit	Credit
3-14	Gas and oil used in company trucks is charged during the month	truck expense	accounts payable
4-10	The gas and oil bill is paid on the 10th of the following month	accounts payable	cash

The 4/10 entry reverses the accrual for accounts payable, so the account now shows a zero balance. A liability account (like accounts payable) increases with a credit, and decreases with a debit. With these two entries, truck expenses are reported in the month incurred. And cash is reduced (credited) in the month actually paid.

You can figure out the appropriate entry for any transaction or accrual by thinking through and identifying the accounts the entry affects.

Here are some typical entries.

	Debit	Credit
cash income	cash	income
accrued income	accounts receivable	income
reversal of accrued income	cash	accounts receivable
cash expense	expense account	cash
accrued expense	expense account	accounts payable
reversal of accrued expense	accounts payable	cash
unearned income	cash	deferred income
reversal, unearned income	deferred income	income
prepaid expense	prepaid expenses	cash
reversal, prepaid expense	expense account	prepaid expenses

The prepaid expenses account is an asset, and any outstanding amounts in that account will appear on the balance sheet. The deferred income is included with liabilities. That way, the balance sheet and profit and loss statement will reflect the true performance of the business.

Balancing the Books

Double-entry bookkeeping has its own built-in error-detecting system. By adding the columns in the journals and ledgers, you can tell immediately if there have been posting or mathematical errors.

At the end of each month, you have to be sure that all entries in the journals are in balance. That is, the debit totals must equal the credit totals. Once you've determined that, you post the general ledger. This process is a simple transfer of totals from the journals to the appropriate pages in the general ledger.

As you can see from Figure 2-3, all journal totals are combined in the general ledger. There, the debit totals should equal the credit totals.

If debits don't equal credits, there's a mistake in one or more of the journals or ledgers. Here are the most common errors to look for:

- An out-of-balance condition in the general ledger from the month before
- A debit posted as a credit, or vice versa
- Transposition (turning numbers around; for example, $315 posted as $351)
- Incorrect addition or subtraction

If the general ledger doesn't balance, follow these steps:

- Recheck the journal balances.
- Recheck the posting of each entry.
- Check math in the general ledger accounts.
- Check the general ledger's balances forward.

Here's a trick for finding a transposing error. If the amount you're out of balance by is a multiple of 9, it's probably a transposition. The sum of the digits in the number by which you are out of balance will add up to 9 or a multiple of 9, and the number itself will be a multiple of 9:

For example, writing $351 instead of $315 will leave you out of balance by $36: 3 + 6 = 9. Posting $288.16 instead of $288.61 will put you put of balance by 45 cents: 4 + 5 = 9.

In the next four chapters we'll discuss each type of journal and the general ledger in detail. Then in Chapter 7 you'll learn how to close the

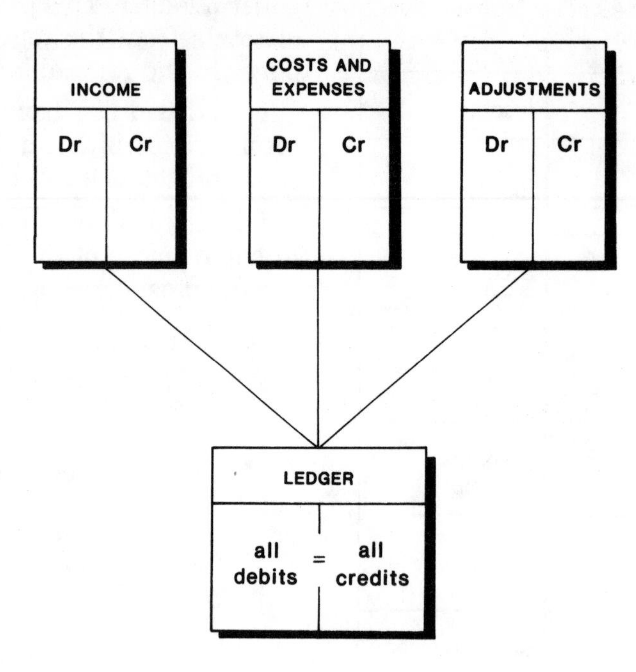

Figure 2-3

books and draw a trial balance (in preparation for financial statements).

We've covered some important subjects in this chapter. But don't be concerned if you haven't mastered the double-entry technique. In later chapters I'll explain how the double-entry system works for each journal and the general ledger. The procedure will become second nature to you as we go on.

But don't leave this chapter until you can answer the questions that follow. If you're unsure of the answer to any question, review this chapter to find the information you missed. Then go on to Chapter 3.

Self Test

1. Cash accounting is a method in which:

a) checking accounts are kept in balance.

b) all transactions are recorded as cash changes hands.

c) income is recorded when earned and expenses are recorded when incurred.

d) none of the above.

2. Accrual accounting is a method in which:

a) income for the year is recorded on the basis of forecast estimates rather than actual receipts.

b) income is recorded when earned, and expenses are recorded when incurred.

c) expenses are deducted only if a job makes a profit.

d) none of the above.

3. During June, a contractor earns $46,800 in income, and incurs costs of $32,100 and expenses of $2,600. Cash receipts for the same month are $33,400; payments total $36,300. On a cash basis, net profit or loss for the month is:

a) $12,100

b) $14,700

c) $(2,900)

d) $(13,400)

4. Using the same facts as in problem 3 above, profit or loss on an accrual basis for the month is:

a) $12,100

b) $14,700

c) $(2,900)

d) $(13,400)

5. A builder earns $5,000 when a job is completed, but isn't paid until the following month. To record the accrued income, the entry is:

a) debit cash, credit income.

b) debit income, credit accounts receivable.

c) debit income, credit cash.

d) debit accounts receivable, credit income

6. Using the same facts as in problem 5, when income is received the following month, the entry to reverse the accrual is:

a) debit cash, credit income.

b) debit cash, credit accounts receivable.

c) debit income, credit accounts receivable.

d) debit income, credit cash.

CHAPTER 3

THE INCOME JOURNAL

Your journals are the books of original entry. You'll recall that these summarize information from source documents. They accumulate the information that you later enter into the general ledger.

Every entry goes through the journal system. This establishes the flow of transactions, and makes the job of posting and balancing the books much easier. Remember, regardless of differences in form, there is only one general ledger. Subsidiary ledgers serve specific functions. And all transactions are posted first to a journal.

The income journal summarizes multiple income transactions. The entries you make to this journal are nothing more than a debit (or series of debits) and a credit (or series of credits). The sum of the debits is equal to the sum of the credits, so total posting adds to zero. This rule is true of all journals. Debits and credits must always equal each other, without exception.

A simple journal entry consists of one debit and one credit. A more involved entry will contain multiple items on one or both sides.

Recording Earned Income

You post receipts in one of two ways. If you're keeping your books on the accrual method (in which you book income when it's earned), you'll make a debit to accounts receivable and a credit to income. Then, when the bill is paid, you'll reverse the accounts receivable entry with a credit, and make a debit to the cash account:

	Debit	Credit
To record earnings:		
Accounts receivable	xxx	
Income		xxx
To record cash receipts:		
Cash	xxx	
Accounts receivable		xxx

the income entry

DATE	ACCOUNT	DEBIT	CREDIT
1-10	accounts receivable	1,000.00	
	income		1,000.00
2-15	cash	1,000.00	
	accounts receivable		1,000.00

Figure 3-1

If you're using the cash method, you'll record income only when the customer pays. In this case, you always debit the cash account, and the credit goes to income. There's no need for a later reversal entry, as no accrual is set up:

	Debit	Credit
To record income:		
Cash	xxx	
Income		xxx

You make a simple income entry for the typical accrual transaction in a general journal. For example, on January 10, you complete a job and bill the customer for $1,000.00 that you've earned. You make an accrual entry for income earned but not yet collected. On February 15, the customer pays the bill. You post a debit to cash and reverse the accrual in accounts receivable. See Figure 3-1 for the actual journal posting.

This is an adequate way to record income when you have only one or two customers. But it's not practical once you exceed this number. Once you move beyond one or two accounts, you should prepare a summary income journal. You simply list all the amounts earned during the month in the summary journal. Then make a single entry to your income journal at the end of the month.

Look at Figure 3-2. This is a summary income journal for a builder who completed six jobs during January. At the end of the month, he made a single entry to the income journal for the $19,600 total of earned income.

Date	Description	Debit	Credit
1-31	Accounts Receivable	19,600	
	Income		19,600

This simple record has several advantages. First, it reduces the number of postings you need to make in a month. This cuts down on the chance for errors, and reduces unnecessary clutter in the general ledger. Second, it summarizes all of your earnings on one page, showing customer names and dates. And, it provides a control for your posting of accounts receivable, which we'll discuss in Chapter 11.

This is only a record of what you *earned*, and not what you actually received in cash. The month-end entry is from the income summary to the income journal to the general ledger. It will consist of a debit to accounts receivable, and a credit to income.

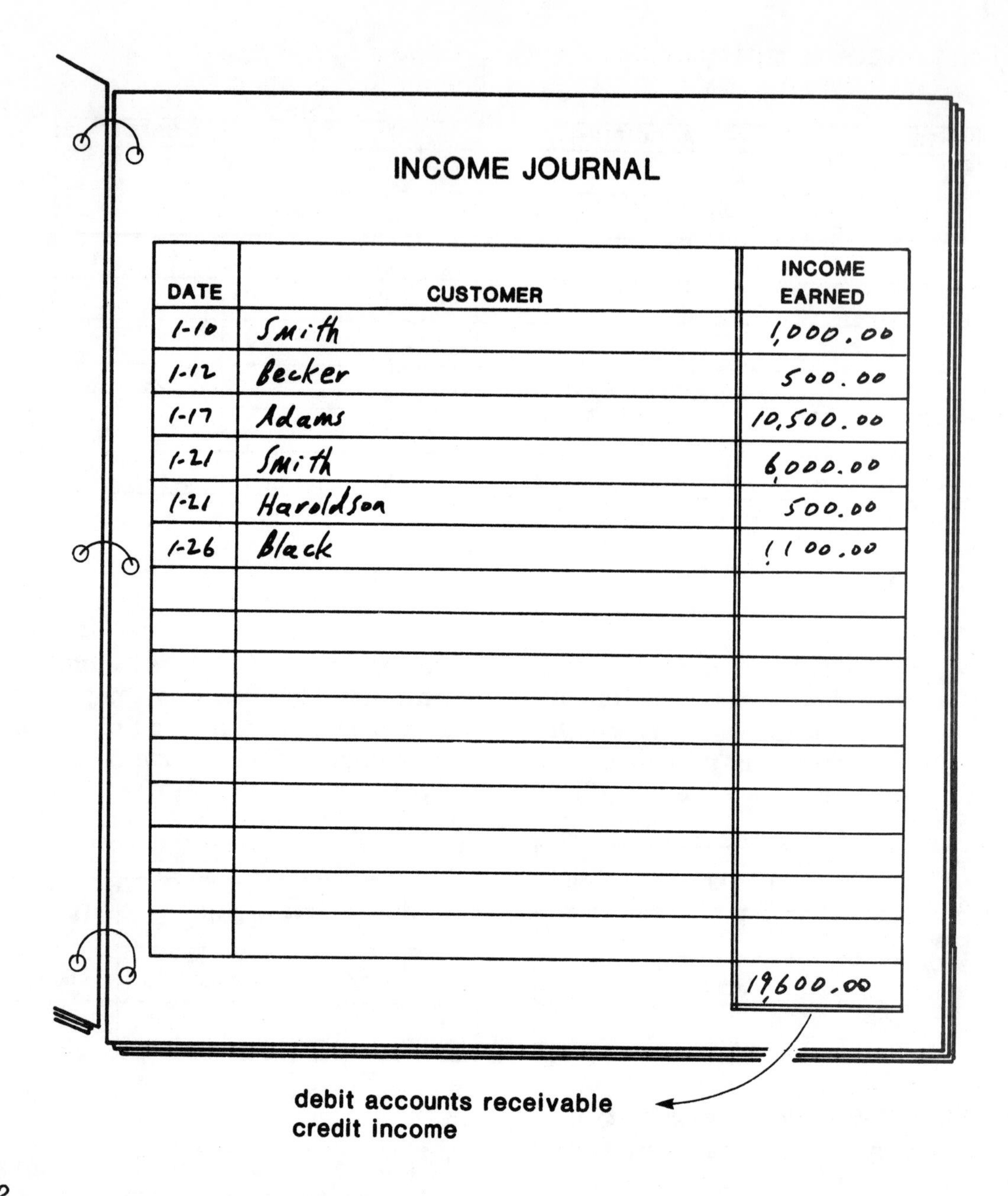

INCOME JOURNAL

DATE	CUSTOMER	INCOME EARNED
1-10	Smith	1,000.00
1-12	Becker	500.00
1-17	Adams	10,500.00
1-21	Smith	6,000.00
1-21	Haroldson	500.00
1-26	Black	1,100.00
		19600.00

Figure 3-2

Balancing Income Records

You need to be sure that this journal has been posted completely and correctly. If you miss recording a job on both the income journal and the accounts receivable ledger, your customer will never be billed.

An easy and consistent way to be sure that you don't miss any items of earned income is to post job cost records and the income journal at the same time. That will give you a way to check your accuracy. We'll cover this in more detail in Chapter 14.

No bookkeeper or bookkeeping system is entirely error-free. You're sure to make some mistakes, and sometimes you'll leave out an entry. The real value of an effective bookkeeping procedure is its ability to catch errors. That way you can make corrections before closing the books each month.

In recording income, you must look out for the accuracy of several different records:

1) The income journal, from which you'll post the general ledger.

2) Accounts receivable subsidiary records, your record of customer billings.

3) Job cost records, where you manage job scheduling and completion billings.

There are ways to avoid making a single entry several times. Certainly, if you keep records by hand, you must make an entry to the income journal, accounts receivable records, and job cost records. That's three different entries each time you earn income. That's time-consuming and adds to the chance for errors.

Some System Alternatives:

1) Automation

In an automated system, a transaction will be posted automatically to all the appropriate records, including the general ledger. In this type of system, you truly enter each transaction only once. There are well-designed computer programs where you make the standard income entry and the debit/credit breakdown is virtually automatic.

Automated systems will make your job easier, once you understand the concepts and procedures of bookkeeping. However, it's a mistake to put the bookkeeping system on computer before you thoroughly understand how to do it by hand. Some contractors automate mainly because they or their employees don't have basic bookkeeping skills. When you do that, you simply turn a bad manual system into a bad automated one. Automate only after you know precisely why and how to keep each type of record by hand.

Automation is also unjustified when you have only a few entries each month.

Efficient management of large amounts of information is what computers do best. So to justify the investment in a computer, you first need the volume that makes it a practical solution. If you don't have that volume, you'll discover that it takes more time to post on the computer than it does by hand. It's also harder to get information out of the computer than out of a hand-posted ledger, unless volume is very high.

2) Write-once systems

You can keep your books on manually posted "peg boards." With this procedure, you make a single entry on a master journal for each type of transaction. At the same time you make the entry, it's recorded through to other ledgers. These systems use either carbon-backed forms or carbonless pressure-sensitive paper. You record a single entry on two records at the same time.

Here's how the system works:

- On the left side of the board, you record income on the individual customer's accounts receivable ledger sheet. This is the permanent record of the customer's account.

- As the information is entered, it's duplicated onto the permanent income journal beneath the ledger card. That way, you post cash receipts at the same time the customer's account card is posted.

With this system, you can verify posting accuracy. The total of postings on the income journal will equal the sum of all income posted to the accounts receivable subsidiary ledger.

Checks and Balances

The way to prevent or find errors is to set up a series of methodical checks and balances as part of your bookkeeping procedure. Effective checks and balances won't increase your bookkeeping load. Rather, they'll reduce the time it'll take you to balance your books. Integrate them into the methods you use to post and record all transactions.

You'll balance records in one book of summary information against detailed record *totals* in another book. Cross-checking will quickly alert you when errors exist, and then help you find and correct them.

Recording Cash Receipts

When you keep your books on the accrual system, there are two elements in your receipts journal. One is for income earned, and the other is for cash received. There's always a delay between billing and payment dates, so the cash receipts in one month will be for income earned earlier.

You enter cash receipts into two accounts. You debit cash (since receipts increase your cash balance) and credit accounts receivable (since you must reduce the amounts owed by customers).

Figure 3-3 shows a cash receipts journal for January. The business received five separate checks from customers, all in payment for income earned (billed) in December. The journal shows the date payment was received, the customer's name, and the amount. At the end of the month, there's one entry to the general ledger showing an increase in cash and a decrease to accounts receivable.

Here, as with the income journal, you need to be sure your posting is complete. If a customer makes a payment that doesn't appear on the cash receipts journal, your books will balance, but they won't be correct. You need to be sure the sum of the entries to your cash receipts journal agrees with your total bank deposits each month. (We'll cover keeping the checkbook in balance later, in Chapter 15.)

An automated or write-once system will minimize the chance that customers' payments will be overlooked. But proper controls of cash, even in a completely hand-kept system, are possible. Control will be simpler if you coordinate your cash control and journal procedures.

The Combined Journal

Both income earned and cash received are related to accounts receivable. You can combine them into a single journal that will reduce the n'·nber and kinds of records you keep. Figure 3-4 shows how you can do this and still keep your journal very simple.

From this single record, you make two sets of entries at the end of the month. For income earned, you debit accounts receivable and credit income. And for cash receipts, you debit cash and credit accounts receivable.

	Debit	Credit
For income earned:		
1-31 Accounts receivable	19,600	
Income		19,600
For cash received:		
1-31 Cash	18,107	
Accounts receivable		18,107

This combined journal lets you monitor accounts receivable records in one place. You can check your posting accuracy in the subsidiary records by comparing their total to the totals for income earned and payments received.

You arrange a journal, like the one shown in Figure 3-4, in order by date. That's because you make the entries to record earned income (billings) or cash receipts when they occur. But you need to file accounts receivable information by customer. And you want to be sure that all entries made on the journal also appear correctly on the customer's ledger.

The change in the total balances due from your customers must equal the change in the income/cash receipts journal. At the end of the month, you add up the ending balance for each customer's account. Compare that to the total from the beginning of the month. The change reflects the addition of income earned, less credits for cash received. That difference should be the same as the difference between the two columns on the income/receipts journal.

Accounts receivable debit	$19,600.00
Accounts receivable credit	18,107.00
Difference	$ 1,493.00

The total of the ending balances for the customer account cards should be $1,493 more than it was at the beginning of the month.

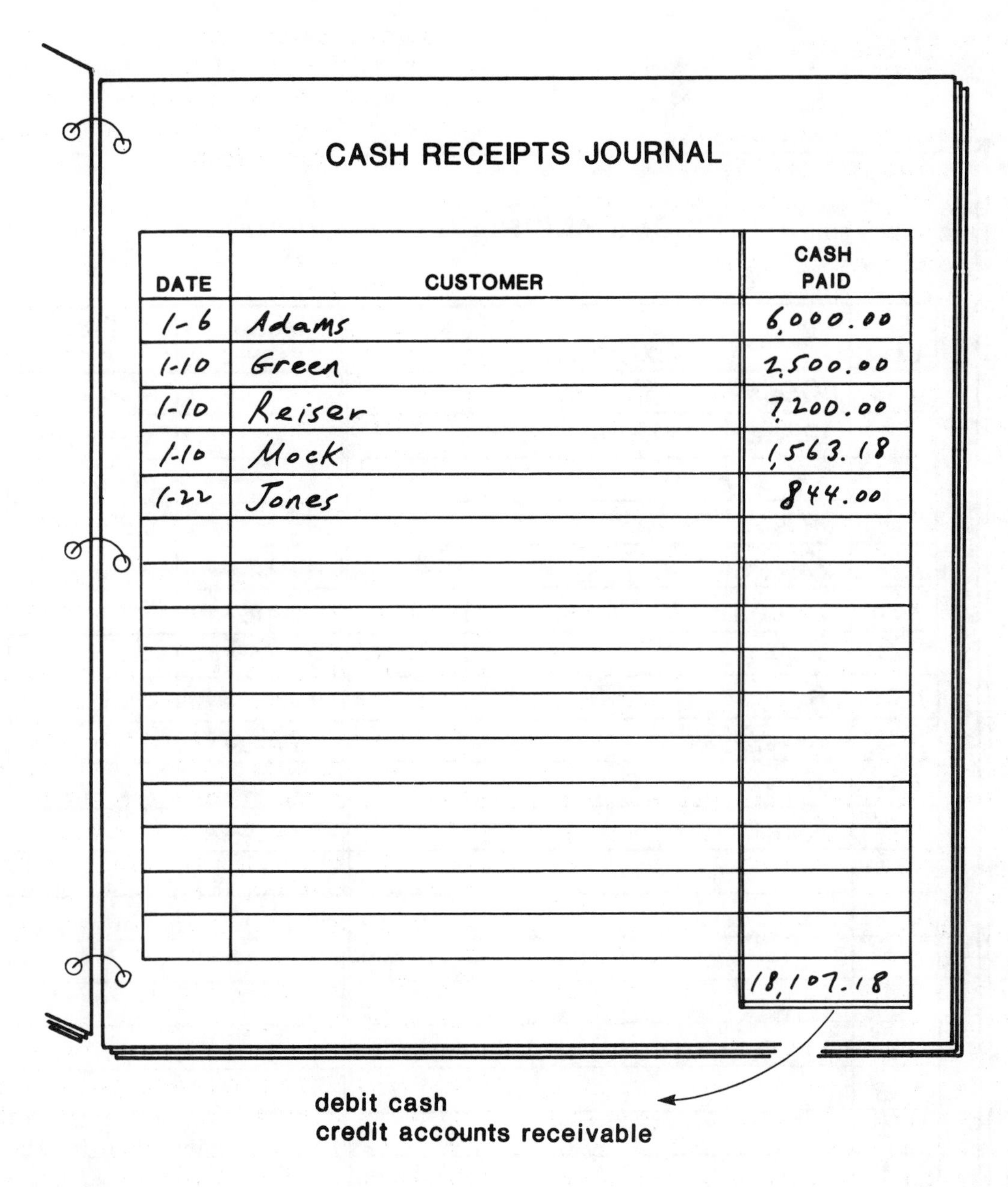

CASH RECEIPTS JOURNAL

DATE	CUSTOMER	CASH PAID
1-6	Adams	6,000.00
1-10	Green	2,500.00
1-10	Reiser	7,200.00
1-10	Mock	1,563.18
1-22	Jones	844.00
		18,107.18

Figure 3-3

In Chapter 11 you'll see how the accounts receivable subsidiary record is balanced and controlled on its own. To make sure that all entries are complete, this cross-checking procedure is a simple but valuable one.

Variations of the Journal

You might want to track your income by different categories. Suppose your income comes from three sources; new home construction, subcontracts on larger developments, and residential repair work. You might want to monitor the relative gross profit on each of these lines of business. (Gross profit is income less direct costs for material and labor.)

You could expand your income/receipts journal to show the following columns to record both earned income and cash receipts.

INCOME AND RECEIPTS JOURNAL

DATE	CUSTOMER	INCOME EARNED	CASH PAID
1-6	Adams		6,000.00
1-10	Green		2,500.00
1-10	Reiser		7,200.00
1-10	Mock		1,563.18
1-10	Smith	1,000.00	
1-12	Becker	500.00	
1-17	Adams	10,500.00	
1-21	Smith	6,000.00	
1-21	Haroldson	500.00	
1-22	Jones		844.00
1-26	Black	1,100.00	
		19,600.00	18,107.18

debit accounts receivable
credit income

debit cash
credit accounts receivable

Figure 3-4

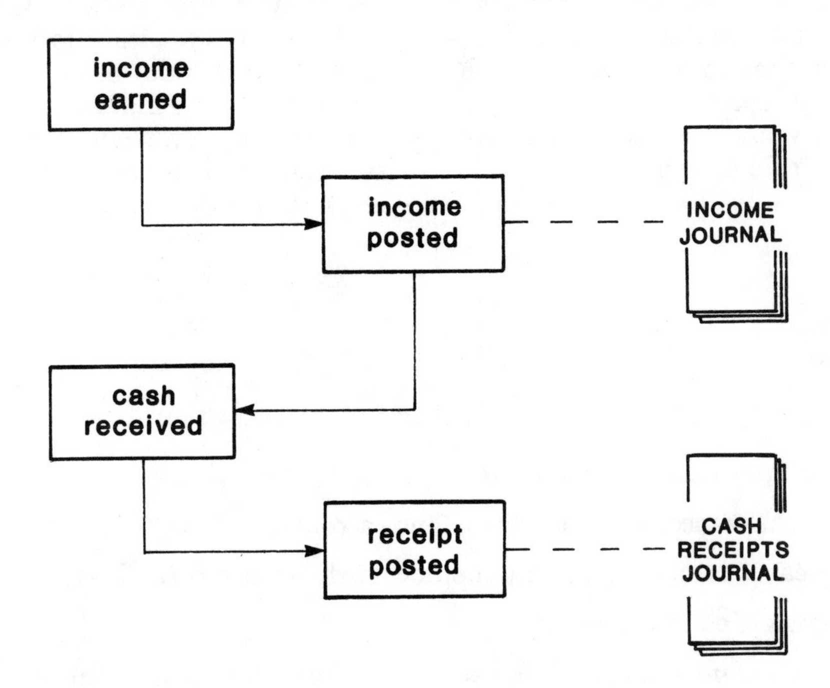

Figure 3-5

Date	Customer	Type of income			Cash receipts
		Homes	Subs	Repairs	

When you bill a job, you'd enter the information in one of the three income type columns. When those are summarized, each part of your business can be posted into a separate general ledger account. When payment is received, you'd enter it under cash receipts regardless of the type of income. This way you haven't expanded the record any more than necessary.

You might be tempted to break income out into many more categories. But remember that the purpose for segregating income is to monitor different lines of business. Don't divide your accounts into more sub-accounts than you actually need. If you don't plan to analyze the breakdown, you don't need it.

Remember that the sequence of recording is identical for all bookkeeping systems, regardless of forms used, or volume:

First: income is earned and posted.

Second: cash is received and posted.

The first step is recorded on an income journal (in isolation or as part of a combined form). And the second step is handled with a cash receipts journal.

All other journals are extensions of this basic form. If you have one receipt per month, a simple two-sided entry is made for income, and another for the cash receipt. When you have a dozen customers, or 50 or 100, you need a more practical summary of transaction volume. But the posting sequence is the same. In Figure 3-5, we've shown the recording flow for income.

You can tailor write-once and automated systems to your specific needs. But be aware that some automated programs are inherently restrictive and inflexible. Be willing to invest enough money to get a program that will suit you. It's a mistake to let program limitations dictate the type of records that you keep.

You can set up manual income journals to suit yourself. You can use a one-column worksheet if you keep separate income and cash receipts journals. You can expand to two columns for a combined journal. And to report on multiple lines of business, you simply use workpaper that provides the number of columns you need.

Stationery stores sell accounting pads with as few as two, or as many as 35 columns. So you have total flexibility to design an effective manual bookkeeping system for your business.

Keep your system as simple as possible, as long as it still gives you the information you need. The more you break down information, the longer it takes to post and manage it. Expand your procedure only when you need the additional information that expansion provides.

Self Test

1. An income journal always consists of:

a) credits only, because income is a credit-balance account.

b) debits only, because under the accrual method, cash is not recorded until later.

c) debits and credits of equal value.

d) debits and credits of unequal value, depending on the change in amounts due from customers.

2. When income is recorded under the cash accounting method:

a) there is no accrual at the time income is earned. It is entered only when cash is received.

b) there is no need to keep records of accounts receivable.

c) you need two journals, one each for earnings and receipts.

d) the income and receipts journals must be kept separate.

3. When income is recorded under the accrual accounting method:

a) an entry is made only when cash is actually received.

b) the income and receipts journals must be kept separate.

c) entries are made to record earnings and to record cash receipts.

d) all income is recorded with a single entry.

4. When you record income and receipts, you must keep track of:

a) changes in accounts receivable balances.

b) cash received and income earned.

c) job cost status.

d) all of the above.

5. *A recordkeeping system expansion:*

a) is necessary whether you have only one receipt per month, or hundreds of accounts.

b) requires automation in order to work.

c) should be dictated by the number of transactions.

d) is best handled with a write-once system.

6. *Recording income earned and cash received on one single form is:*

a) a practical record that also helps control accounts receivable records.

b) a poor idea, since combining two different records will lead to many errors.

c) only possible with a fully automated system.

d) acceptable if you have only one or two customers.

CHAPTER 4

THE PAYMENTS JOURNAL

The last chapter discussed records for control of income. Now we'll talk about the second major type of journal which deals with expenses and costs — money paid out.

It's important to know not only what you spend during the month, but also what you owe. When you order materials, you'll keep a record so you'll know how much debt you incurred during the month.

What Payment Records Should Do

You need to keep complete records to be sure that you don't overlook any costs and expenses. You also want your accounts payable records to be accurate and up to date, even when your suppliers haven't billed you yet. And if you're on an accrual system, you'll want to know which payments are for current expenses, and which are for costs incurred earlier.

Your bookkeeping procedure should let you do the following:

- Identify and pay accounts payable on time.
- Reconcile those payments against your records of money owed.
- Know how much you owe in the future for purchases you make this month.
- Keep your checkbook in balance.
- Keep consistent records of cash payments.

Let's start with accounts payable. On an accrual system, you'll end each month by making a list of what you owe to other people. Then you enter that liability into the books.

The next month you'll reverse the accrued accounts payable in one of two ways. You'll either make a reversing journal entry, or you'll write checks against your accounts payable balance.

First, let's consider the journal entry method. At the end of the month your journal entry for total materials purchased looks like this:

Date		Debit	Credit
4-30	Materials	10,081.27	
	Accounts payable		10,081.27

Your balance sheet and profit and loss statement will accurately show what your expenses were for the month, and how much you owe. After you close the books for the month, you reverse the entry in the journal this way:

Date		Debit	Credit
5-1	Accounts payable	10,081.27	
	Materials		10,081.27

Now, when you write checks during May to pay bills from suppliers, you'll code the checks to materials.

"Coding" refers to a number that you assign to each account in your general ledger. We'll discuss how to assign account codes in detail in Chapter 17. The reason for codes is so you don't have to write the complete account name every time you post a journal entry. Numbered account codes are necessary if and when you automate your bookkeeping procedure.

If you *don't* reverse the journal entry, you'll code the checks to accounts payable. In the above example, the materials account was debited in April when the orders were placed. In May, the bills are paid with checks coded to reduce the accounts payable balance.

The problem with this method is that it's hard to distinguish which costs and expenses were accrued during a previous month and which are for current month purchases.

You'd have to check each unpaid bill against the list you made at the end of the previous month. Then you'd have to be sure the bill was coded properly when you paid it. Each month, you'd have to reconcile your entries to be sure that all the accruals were reversed.

It's much simpler to reverse the entire entry which also reduces the cost account balances. Then increase those account balances again when you write the checks.

Tracking an Accounts Payable Transaction

For the moment, let's assume you agree that the best procedure is to record and reverse accruals by way of a journal, and assign costs and expenses to the proper account when you pay them.

Suppose you owe $4,715.80 for materials at the end of April. You'll pay these bills in May, when you write checks to your suppliers. These are the steps you'll take:

- On April 30, you prepare an accrual journal entry showing a debit to materials and a credit to accounts payable.
- In May, you prepare a reversal journal. It's exactly the opposite of the April 30 entry.
- During May, you write checks totaling $4,715.80, and code them to materials.

The two journal entries are shown in Figure 4-1. As you see, they offset one another. But when you look at the summary of these entries, you'll see that all entries — including materials, cash and payables — end up in the right month. The purpose of accruing is to post materials in the month they are used, even when payment lags one month behind. A summary of these entries:

	April		May	
	Debit	Credit	Debit	Credit
Materials	4,715.80		4,715.80	4,715.80
Payables		4,715.80	4,715.80	
Cash				4,715.80

the payment entry

DATE	ACCOUNT	DEBIT	CREDIT
4-30	Materials	4,715.80	
	Accounts Payable		4,715.80
5-1	Accounts Payable	4,715.80	
	Materials		4,715.80
5-31	Materials	4,715.80	
	CASH		4,715.80

Figure 4-1

An alternative is to separate your bills between accounts payable and current month expenses. I don't recommend this. It causes extra work and makes it harder to keep your accounts payable account in balance.

Here's what can happen. Suppose you maintain two checkbooks, one for accounts payable and the other for current bills. If you pay a current bill out of the payables account, your payables won't balance. The total of the checks written for accounts payable will include an amount that wasn't posted to accounts payable in the first place. Also, the purchase won't be posted to the appropriate cost or expense account. You'll have to go back later and make a correction.

If you keep the accounts payable checkbook strictly to pay month-end liabilities, you must not use that account for any other purpose. But it's easy to make a mistake in the pressure and rush of the daily routine. When you write a check from the wrong account, you'll only make your job harder. It's simply easier to use journals for accruals, and then to write all checks from one checkbook and post them to the appropriate cost or expense account as they're paid.

The Cash Disbursements Journal

Just as the income journal in the last chapter is a specialized record to summarize cash receipts and accounts receivable, the cash disbursements journal records your payments.

Most active businesses write many checks every month. So it wouldn't be practical to post each one to the general ledger individually. That would take too much time and would clutter your general ledger with too many entries. The alternative is to use one cash disbursements journal, like the one in Figure 4-2.

Design your journal to suit your own operation. For example, if you make frequent payments for truck expense, that should have its own column. Set up the journal to distribute those expense

CASH DISBURSEMENTS JOURNAL

DATE	PAID TO:	CHECK NO.	TOTAL	MATERIALS	LABOR	PAYROLL TAXES	SUPPLIES	OTHER PAYMENTS	
								AMOUNT	ACCOUNT
2-4	James Management	3218	950.00					950.00	Rent
2-5	Payroll account	3219	8,415.00		7,550.00	865.00			
2-8	Southern Supply	3220	234.16				234.16		
2-8	VOID	3221	-0-						
2-10	Harvey & Hanks	3222	2,480.00	2,480.00					
2-10	Carson Lumber	3223	1,506.00	1,506.00					
2-10	Wilson Supply	3224	387.60	387.60					
2-10	Peterson Supply	3225	4,144.55	4,144.55					
2-12	Payroll account	3226	8,217.35		7,402.85	814.50			
2-12	Central Phone Co.	3227	186.40					186.40	Telephone
2-19	Payroll account	3228	8,340.00		7,497.00	843.00			
2-22	Lake office Supply	3229	432.77				432.77		
2-24	Baskin Auto	3230	255.00					255.00	Truck Expense
2-26	Payroll account	3231	8,606.15		7,733.35	872.80			
2-26	Todd Lumber	3232	1,580.00	1,580.00					
2-26	Postmaster	3233	66.00					66.00	Postage
			45,800.98	10,098.15	30,183.20	3395.30	666.93	1,457.40	

credit cash

debit various accounts

Figure 4-2

accounts you use a lot. Then use the last two columns for "all other."

The disbursements journal gives you many forms of control:

- It accounts for every check. (Always use numbered checks and post the journal in order by check number.)
- It gives you a record of all payments by date and check number.
- It breaks down all payments into their appropriate accounts, so it's easy to make one month-end entry.
- It allows you to balance cash *before* you post to the general ledger.

This last point is critical. Don't post from a journal to the general ledger until you've checked the math accuracy in the journal. Notice that every payment is entered twice. One entry is for the total (which is a reduction, or credit to cash), and the other for the distribution (various debits).

Before you post this journal to the general ledger, add the column totals to make certain that the total of the distribution columns equals the "total" column.

After checking the math, you can post all of your monthly payments in one easy step. Figure 4-2 will be posted like this:

Date	Description	Debit	Credit
2-28	Materials	10,098.15	
	Labor	30,183.20	
	Payroll taxes	3,395.30	
	Supplies	666.93	
	Rent	950.00	
	Telephone	186.40	
	Truck expense	255.00	
	Postage	66.00	
	Cash		45,800.98

This is a summary of the debits and credits you'll make from the cash disbursements journal. It's a more accurate control and is easier to manage than an entry for every payment. And the entire month, regardless of how many checks you write, is reduced to one entry per account.

The Purchases Journal

Simplifying the cash disbursements journal, and handling accruals through a once-monthly general journal entry, is practical when you have a low volume of business. But when volume is high, you need a more detailed procedure.

You'll have trouble keeping track of job costs if you don't post orders for materials correctly, or if they're not posted to the books until the month after they're used.

One solution is a requisitioning system. When materials are needed, the job supervisor fills out a numbered materials requisition form. The requisition goes to the bookkeeper where it supplies several valuable pieces of information:

- It furnishes figures needed to set up an accounts payable accrual.
- It provides information for job cost records. (This is a separate subsidiary record, described in detail in Chapter 14.)
- It gives you information you need to follow up on orders, and it's numbered for reference and posting control.

This is similar to the control over checks in the disbursements journal, where every number can be accounted for. The difference is that not every requisition will be used in order, and some might even be destroyed and not used at all. But if you have a requisition system, you won't be as likely to duplicate or miss recording a payable.

Sometimes you can't control requisitions. This system won't work as well as it should if you have two or more foremen who keep requisitions in their truck and use them as needed. The requisition system is most practical when you can control the distribution and use of requisition forms.

PURCHASES JOURNAL

DATE	ORDER NUMBER	VENDOR	TOTAL
4-15	405	North Bay Supply	4715.80
4-15	406	Matson Lumber	2,115.00
4-15	409	Beech Hardware	307.82
4-21	416	Sampson & Sons	318.60
4-23	417	Matson Lumber	1,404.55
4-23	419	Teague, Inc.	916.00
4-29	424	Wentley Supply	303.50
			10,081.27

debit materials
credit accounts payable

Figure 4-3

Another alternative is to use an unnumbered form and assign a current number to it when it enters the bookkeeping system. This won't prevent a completed form from being lost before it gets to the bookkeeper. But at least you'll be able to fully account for all the paperwork you do receive.

Here's how the requisition procedure works using a three-part form:

- The supervisor or foreman orders materials at the job site by filling out the requisition. A copy of the requisition is kept in a file at the site.
- The business owner approves the requisition, and forwards the remaining copies to the bookkeeper.
- The original is sent to the supplier. The first copy is filed as the permanent bookkeeping record copy.
- Once a week, the bookkeeper posts the purchases journal, using the permanent bookkeeping record copy of the requisition as the source document. Figure 4-3 is a sample purchases journal.

ACCOUNTS PAYABLE JOURNAL

DATE	PAID TO:	CHECK NO.	ACCOUNTS PAYABLE	DISCOUNT	NET CHECK
5-10	North Bay Supply	3775	4,715.80	235.79	4,480.01
5-10	Matson Lumber	3776	3,519.55	70.39	3,449.16
5-10	Teague Inc.	3777	916.00	18.32	897.68
5-10	Beech Hardware	3778	307.82	6.16	301.66
5-16	Wentley Supply	3779	303.50	–	303.50
5-17	Sampson & Sons	3780	318.60	–	318.60
			10,081.27	330.66	9,750.61

debit accounts payable

credit discounts earned

credit cash

Figure 4-4

At the end of the month, you'd post the total for purchases into the books. You'd increase (debit) material purchases and credit the accounts payable account.

The *amount* entered on the purchases journal might be your best estimate rather than an exact price. Some suppliers offer discounts for prompt payment or quantity purchases. Prices might change. Or some part of a total order might be out of stock and back-ordered. In any of these cases, the amount recorded on the purchases journal won't be correct. That means you'll have to make an adjustment in the following month.

The adjustment will be made when the bills are received and paid. In an ideal situation, you accrue exactly the amount you will owe, all materials come in on time, and invoices are received and paid on time. In practice, this will rarely occur.

In Figure 4-4, we show how a "perfect" reversal can be made with the accounts payable journal. Here, all of the entries in the previous month's purchases journal were paid, and the accounts payable entry is completely reversed.

Notice that the actual cash paid out is less than the accounts payable. This is due to discounts the suppliers allowed in some cases. That's one reason adjustments are needed.

More likely causes are different prices or partial filling of orders. When prices are different, you'll have to adjust accounts payable and material entries with an adjustment journal.

Let's look at Figure 4-4 again. What if the last payment, to Sampson and Sons, was actually for $345.80, or $27.20 more than was posted in the purchases journal? If this happens, the debit to accounts payable is higher than the accrual in the previous month. That means that the materials account is also wrong. An adjusting journal is necessary.

Date	Description	Debit	Credit
5-17	Accounts payable	27.20	
	Materials		27.20

As you can see, this would be a time-consuming process if you had to do it for each invoice that didn't agree with the purchases journal entry. You can avoid making these individual adjustments if you use the purchases journal this way:

- Use the requisition system to document materials ordered, to post the purchases journal, and to set up a month-end accrual.
- Reverse the whole accrual with a single entry in the general journal.
- Make current payments by coding to the appropriate cost account, not accounts payable.

This way you still have control over materials. You can still post the job cost records when you place orders. And you can still set up accrued expenses at the time you close the books. But you don't have the on-going problem of balancing accrued accounts payable to actual payments. You reverse the accrual in full each month.

Managing the Checkbook

Remember that recording checks is part of the second phase of recordkeeping. The first is the source document — an invoice, receipt or voucher for example. Second is the book of original entry, in this case the checkbook that will eventually be summarized in a disbursements journal. The third will be the general ledger.

When you record payments properly, you'll be sure all the bills are paid, and also make sure that all the proper accounts have been posted.

Figure 4-5 shows a page from a typical hand-kept business checkbook. These stubs show the balance forward, a reduction for each check written, and an ending balance. You'd add any deposits. Each stub shows the date, the payee, and a description of the payment. The first check includes a discount, so the coding is broken down between accounts payable and discounts.

The checkbook serves several bookkeeping purposes. First, it shows all the cash activity. Second, it's a source for information to post. And third, it lets you know your cash balance.

To check the accuracy of the math on a check stub of this type, start with the balance on the bottom of the page, and add each of the three checks. If your total equals the balance on the top of the page, you know your math was correct. When there are deposits, subtract those from your total.

You can use the same balancing procedure whether you keep a running total on your check stubs or use another system. For example, a pegboard format produces a check with posting information arranged on one line. The check is designed so the date, payee, check amount and distribution accounts are all copied onto a journal sheet as you write the checks. You should total the disbursements each day and transfer them to a cash balancing worksheet along with a summary of the day's deposits.

You'd still start with the ending balance, add all checks and subtract deposits to arrive at the previous day's ending balance.

the checkbook

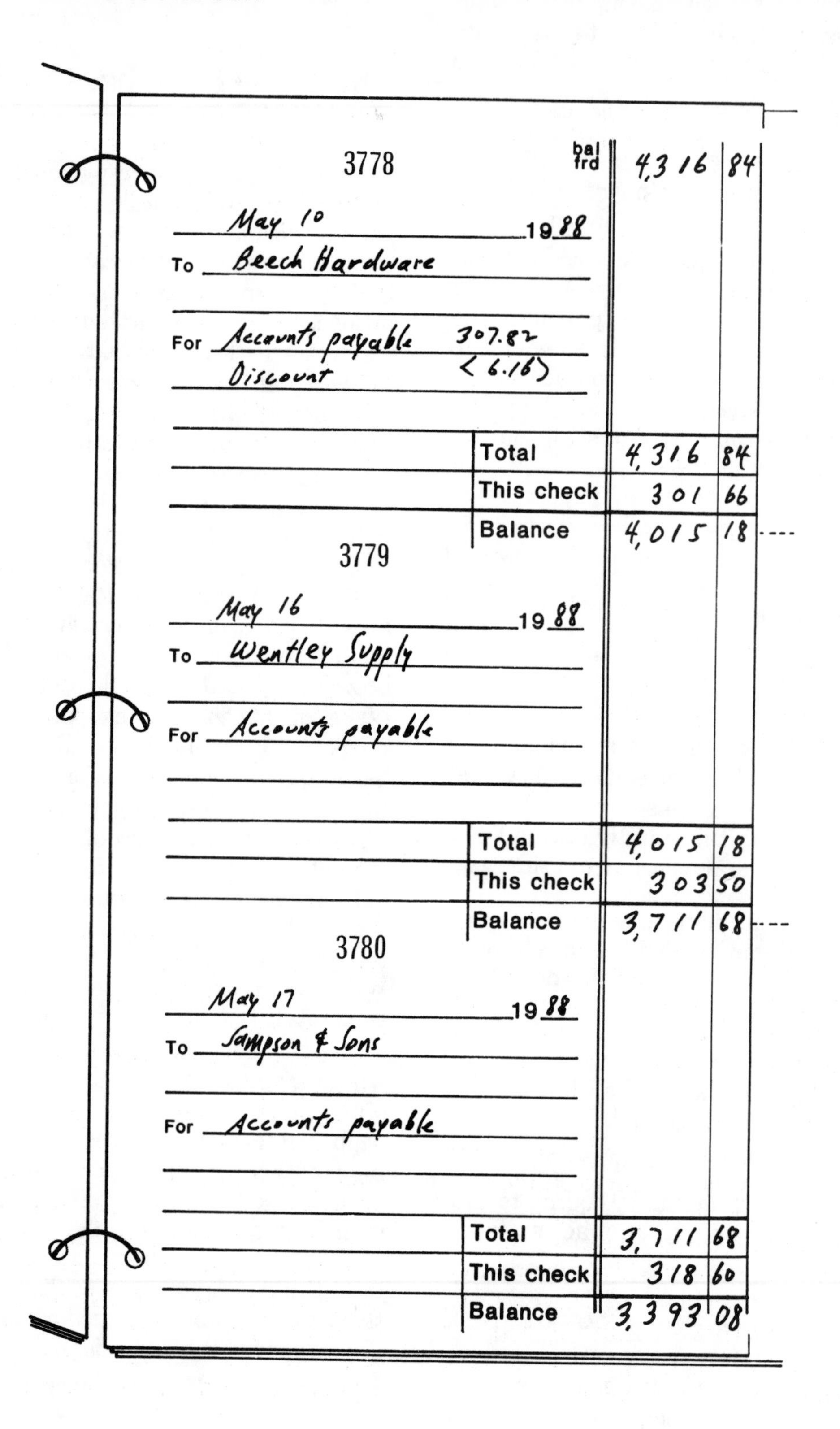

Figure 4-5

It's probably not efficient to post the cash disbursements journal every time you write a check. So an accurate and complete checkbook is essential to keeping good books.

You'll probably make many payments at one time, and then set aside a block of time at the end of each week to post all of your books at once. You could even do this once a month. But if you've left out important information, it will be harder to post your disbursements journal. So both for cash control and bookkeeping efficiency, keep the checkbook as complete as possible. That is an essential first step.

Are Two Checkbooks Better Than One?

When should you have multiple accounts? There are some good reasons, but remember that when you keep two checkbooks, you add to the month-end balancing process. You can justify multiple accounts only when:

- The second account saves time or simplifies your bookkeeping task.
- You need separate records more than you need simplicity.
- You can save for future expenses with a separate fund.

You might consider a separate payroll account. It may be necessary when your payroll is done by an outside service company or by a bank. The bank prepares the detailed records and tax returns and writes the checks. You would then transfer the money needed for the payroll and taxes from your general account to the payroll account on payday.

You might also want to use a special account to reserve money for other taxes. If you have to make periodic payments for payroll, highway use, gasoline, excise, or income tax deposits, you can use a special tax account as a reserve. You figure how much you need weekly, even though the full amount isn't payable until the end of the quarter. You deposit that amount into the tax account each week. This way, the burden of large future payments doesn't hit all at once.

As I suggested earlier, you might want a separate checking account for accounts payable. Keep in mind that this complicates keeping the books in balance. You'd also have to transfer money back and forth between accounts — an added nuisance.

As an alternative, you can pay accounts payable and current expenses from the same account, but use two different check formats and numbering systems. This complicates your monthly bank account reconciliation, since you have to account for two sets of outstanding checks instead of one. If you don't mind the inconvenience, this might be your best solution.

Control over cash accounts is explained more fully in later chapters, especially Chapter 9 (payroll records), Chapter 12 (records for payables and purchases), and Chapter 15 (keeping the checkbook in balance).

The Flow of Monthly Payments

In the ideal situation, all your accrued expenses will be exactly right and will be reversed in full the next month.

Figure 4-6 shows how the flow of payments works. Each month's purchases journal is used to set up the accrual. This is reversed each month as payments are made. Unpaid accounts are carried over to future months. At the same time, current expenses are paid from the cash disbursements journal.

Consider modifying this system slightly. Keep the purchase requisition and the purchases journal each month. Use this to post job cost records and to prepare your accrual. But then reverse the accrual entirely as soon as you start a new month, and make current payments against the appropriate material purchases account.

This way, you won't carry over unpaid amounts as accounts payable. Anything still owed at the end of each month is set up again as an account payable, but the entire amount is eliminated when you open the next month. It's simple, easy to balance, and it does away with the need for constant and time-consuming adjustments.

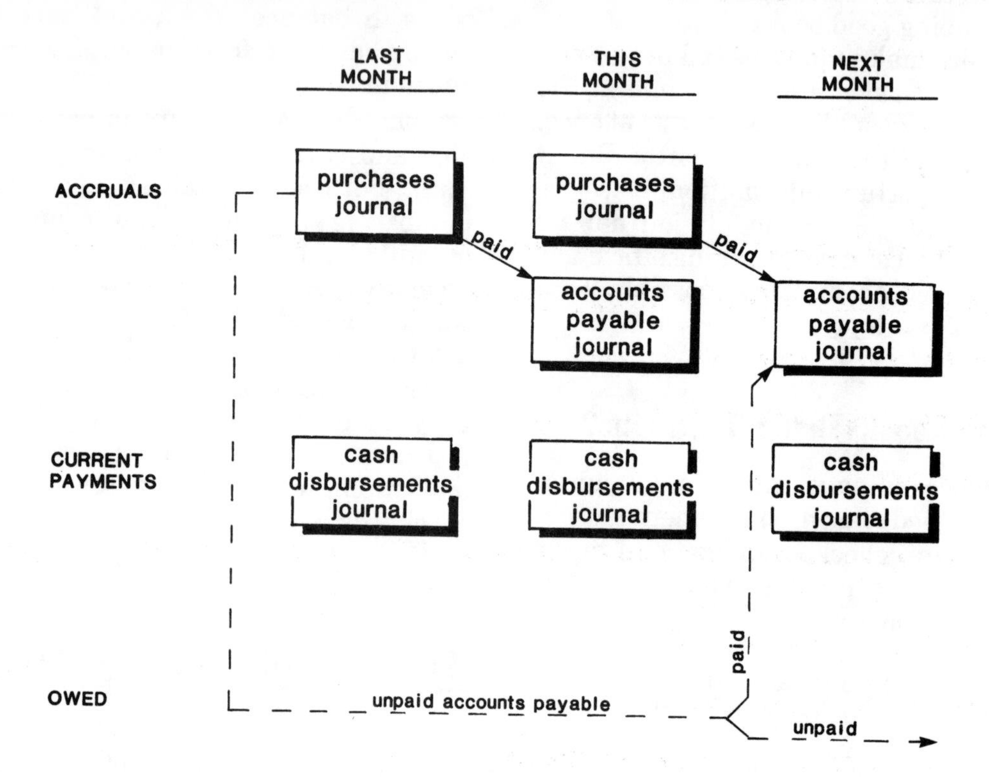

Figure 4-6

Self Test

1. The purpose of accruing accounts payable at the end of the month is to:

a) recognize costs during the proper month.

b) account for all outstanding requisitions.

c) offset entries to earlier accounts payable.

d) all of the above.

2. A requisitioning system helps to:

a) avoid duplicate payments.

b) keep track of accounts payable.

c) post job cost records currently.

d) all of the above.

3. *A disbursements journal provides many forms of control, including:*

a) reconciliation of accounts payable.

b) math accuracy before posting the ledger.

c) daily cash balancing.

d) all of the above.

4. *The problem with a separate bank account for accounts payable is that:*

a) you must reconcile payables to actual payments each month.

b) you have two bank accounts to balance.

c) you could double-post an account by mistake.

d) all of the above.

5. *When you accrue accounts payable at the end of the month, then make a reversing entry, the effect is to:*

a) inaccurately post materials twice.

b) make extra work for yourself.

c) recognize all payments in the proper month.

d) none of the above.

6. *You should have two separate checking accounts when:*

a) the volume of monthly payments makes it practical.

b) you have a complete requisitioning system that is balanced each month.

c) keeping checking accounts in control is not a problem.

d) all of the above.

CHAPTER 5

THE GENERAL JOURNAL

In the last two chapters we described how receipts and payments are handled on specialized journals. All other types of entries to the general ledger come from the general journal. Those include monthly adjustments, accruals and reversals of previous accruals, corrections, non-cash expenses and any other entry that is not a receipt or a payment.

The Accrual Journal

The format of the general journal is simple. Two columns on the far right side of the page contain debits and credits. And like all types of entries, these always equal one another. To the left of these columns is a narrow space where you check off each entry as it's recorded in the general ledger. Left of that is an explanation area, where you'd write either the account name or a coded description. And to the far left is the date:

Date	Description	√	Debit	Credit

Each journal entry is numbered, either in sequence, or by month. For example, you can number the first entry in February as 2-1. Finally, you should always include an explanation at the end of the journal entry, in the explanation column.

Make the explanation brief but completely descriptive. Some examples of explanations:

1) To accrue accounts payable at 2/28
2) To reverse 2/28 accrual of accounts payable
3) To adjust cash balance per bank reconciliation
4) To correct miscoded check #2313

Notice that each of these explanations tells exactly why the journal entry was made. If an auditor wants to check further, the explanation leads to the right place. Items 1) and 2) above can be traced to the accounts payable file. You

the accrual journal

DATE	EXPLANATION	✓	DEBIT	CREDIT
	-7-			
7-31	Materials		8,475.91	
	Travel Expenses		316.00	
	Office Supplies		107.45	
	Telephone		218.40	
	Utilities		107.63	
	Accounts Payable			9,225.39
	to record accounts payable			
	as of 7-31			

Figure 5-1

can check the bank reconciliation for Item 3), and the coding change refers to the checkbook.

Never prepare a journal entry without an explanation. What's clear and logical to you today might not make any sense at all later on. If you're ever asked why you made a particular journal entry, your explanation wasn't clear enough.

The explanation justifies every type of entry to the general journal. The first type you'll prepare is for accruals. If you post accruals to your books each month, you'll post a debit to various cost and expense accounts, and a credit to accounts payable.

You'll reverse the accounts payable entry in the following month, in one of two ways. The easiest and clearest way is to prepare a reversing journal entry. This way, the entry goes in one month and out the next. You don't have to reconcile the accounts payable ledger with the actual bills outstanding.

If you have a large volume of accounts payable, the accrual journal is reversed when you pay the bills during the following month. When you write the checks, you code them as debits to accounts payable. These debits offset the journal entry credit you made at the end of the previous month.

I recommend the entry and reversal method for most small and medium-sized contracting firms. In the last chapter, we showed how an accounts payable checkbook system works for large-volume business. But most businesses can work out of a single checkbook, and use the general journal for month-to-month accruals.

Figure 5-1 shows you a journal entry for the accrual of accounts payable on July 31. The entries are a summary of a list of bills outstanding at month end.

The Reversal Journal

Whenever you prepare a journal that accrues income or expenses, you have to reverse the entry in some way. Remember that the accrual

the reversal journal

DATE	EXPLANATION	✓	DEBIT	CREDIT
	-1-			
8-1	Accounts Payable		9,225.39	
	Materials			8,475.91
	Travel Expenses			316.00
	Office Supplies			107.45
	Telephone			218.40
	Utilities			107.63
	to reverse accounts payable			
	accrual of 7-31			

Figure 5-2

is an entry to recognize something at the end of a reporting period. You remove it as cash transactions occur.

It may seem like a lot of extra work to make the entries one day and reverse them the next. But when you consider how accruals affect the financial statement, you can see their value and importance.

If you look again at Figure 5-1, you'll see that the accounts payable total is $9,225.39 at the end of July. Before posting those liabilities, the general ledger shows a profit of $6,000. But in fact, the year-to-date loss is $3,225,39. Without the accruals, the financial statement is seriously misleading.

The solution is to accrue the liabilities and then prepare a completely accurate financial statement. But because you'll probably pay those debts in August, you need to take the accrual back out of the books for two reasons:

1) The liabilities will be paid and will no longer be owed. If you leave them on the books after you pay them, the financial statement will be inaccurate again by showing the same costs twice — once as an accounts payable liability and again as an expense.

2) At the end of August you'll have another group of unpaid bills that will be due in September.

First you have to reverse the previous liability, and then accrue the new one. A journal entry to reverse an accrual is exactly the opposite of the accrual. This is where the general journal simplifies your life. You just turn the entry around, debiting accounts payable and crediting the various accounts in the accrual entry from the month before. You can see this in Figure 5-2.

Adjustment Journal

Besides accruals and reversals, the general journal is used for adjustments and corrections. For example:

- You discover errors in coding of checks from the previous month.

- As part of the bank reconciliation, you need to record service charges and the cost of having checks printed.

Here are some adjustments you'd need to enter after your bank statement comes:

- Service charge -6.00

- A charge for printed checks -28.50

- One returned check -250.00

- Redeposit of previously returned check +100.00

You can summarize all these adjustments in a single journal entry because they all affect a common account, cash.

- The service charge, printed check charge, and returned check all reduce the cash balance.

- The redeposit is an increase to cash.

The entry to cash is a credit of $184.50. (Remember, the cash account increases on the debit side.) You can check this in two ways. First, that's the net of the three reductions, less the one increase. Second, this amount will agree with your bank reconciliation (adjustments, excluding outstanding checks and deposits in transit).

So you'll credit cash for $184.50. The sum of your debits must also total that amount.

The service charge of $6.00 should be a debit to an expense account, Bank Charges. The charge for printed checks is usually treated as Office Supplies, another debit.

The other two entries are adjustments to the Accounts Receivable account. The returned check is a debit because it increases accounts receivable. This was previously assumed to be paid, when the check was received. At that time, you debited cash and credited Accounts Receivable. Now, the check has been returned, meaning the customer owes you the money again.

The redeposited check for $100 is the opposite, a debit to Accounts Receivable. This is the third part of a total entry, going back to the original payment:

	Debit	Credit
Original Payment:		
Cash	100.00	
Accounts Receivable		100.00
Returned check:		
Accounts Receivable	100.00	
Cash		100.00
Redeposit:		
Cash	100.00	
Accounts Receivable		100.00

So the total details of this entry would be shown in an adjusting general journal as:

Date	Description	Debit	Credit
	Accounts Receivable	250.00	
	Accounts Receivable		100.00
	Office Supplies	28.50	
	Bank Charges	6.00	
	Cash		184.50
	-to record adjustments to cash, per bank reconciliation, including redeposit of $250 check from Jones; returned check for $100 from Brown; printed checks and bank charges-		

While accruals and reversals offset one another, the adjustment journal is permanent. You

adjustment journals

DATE	EXPLANATION	✓	DEBIT	CREDIT
	-1-			
8-4	Office Supplies		35.00	
	Postage			35.00
	to correct miscoding on			
	check #7107			
	-2-			
8-5	Cash		1,206.17	
	Bank Charges		8.00	
	Miscellaneous		1.43	
	Materials			1,215.60
	to record adjustments to			
	cash, including voiding of			
	check 6415, for materials			

Figure 5-3

are correcting or adjusting account balances, so you don't reverse them later. A typical adjustment journal is shown in Figure 5-3.

The Non-cash Expense Journal

You write checks for most business expenses, and record them on the payments journal (see the previous chapter). But some types of expenses must be recorded through the general journal.

Here's an example: Your foremen regularly ask for reimbursement of cash expenses, giving you receipts for phone calls, car expenses, and other minor payments. Often these payments are too small to justify writing a check for each one, so you pay them out of your own pocket.

At the end of the month, you add up the total of these receipts and enter them through the journal. The entry would consist of a debit or series of debits to the proper expense accounts, and a credit to your capital (equity) account since you're not being reimbursed in cash. (You've invested your own money into the business.)

Be sure you're consistent and thorough in collecting the receipts for reimbursements of this kind. If you overlook expenses or don't keep the source documents and want to capture the expenses at the end of the year, you might try to estimate the totals. That makes you very vulnerable in case of a tax audit.

One contractor we know paid for fill-ups for employees' personal cars out of his pocket, but kept no records. When it came time to close the books, he estimated the total and made a journal entry. But his tax return was audited, and the IRS disallowed the entire deduction. Why? Because there was no documentation for the expenses. Estimates were not good enough.

A better solution is to use a petty cash fund, with enough money in it to cover cash expenses for a month. We'll deal with this in more detail in Chapter 13.

In the case of some payments, your checkbook can't tell the whole story. If you make a deposit on a new piece of equipment, you enter it into your checkbook as payment for a fixed asset. The equipment is actually worth much more than the amount of the check, and the transaction includes a liability. For example, putting $1,000 down on a $10,000 purchase, the entry should read:

Date	Description	Debit	Credit
	Fixed Assets	10,000.00	
	Notes Payable		9,000.00
	Cash (check ####)		1,000.00
	-to record down payment on (fork lift, for instance)-		

You'll have a number of expenses that you must enter through the general journal. Not all of your expenses are paid by check or from petty cash.

Here are some examples of non-cash expenses that occur monthly:

1) Amortizing a prepaid asset

You can amortize (spread) a one-time payment for insurance over the term of the policy. If you pay in advance for a two-year policy, you should book 1/24th of the cost to expense each month.

At the time the payment is made, it's recorded as a prepaid asset:

Date	Description	Debit	Credit
	Prepaid Assets	4,632.00	
	Cash		4,632.00

Then, 1/24th, or $193 is reversed out of Prepaid Assets and recorded to Insurance Expense each month, by way of a non-cash expense journal:

Date	Description	Debit	Credit
	Insurance	193.00	
	Prepaid Assets		193.00

At the end of 24 months, the Prepaid Asset account will be amortized down to zero, and each month's insurance expense will show a debit of $193.

2) Amortizing organizational expense

Another form of amortization has to do with startup expenses. When a business first opens, some money is spent that is properly called organizational rather than operational. For example, when you first open your doors, you pay an attorney to negotiate a lease. You buy fixtures for the office and pay general overhead expenses for three months before actually opening for business. You might also pay an accountant to set up your books.

All of these are organizational expenses that are set up as assets and amortized. In most cases, you'd write these off over five years. At the time you start your operation, the organizational expenses are set up as assets:

Date	Description	Debit	Credit
	Organizational Expenses	4,707.00	
	Capital		4,707.00

This is how much you spent to get the business going. But now, you want to amortize 1/60th of that total each month. This is a non-cash expense, since the money was spent before you even opened your doors:

Date	Description	Debit	Credit
	Amortization	78.45	
	Organizational Expenses		78.45

At the end of 60 months, the asset, Organizational Expenses, will be at zero, and each month will show $78.45 in amortized expenses.

3) Depreciation

You should record depreciation whenever you prepare a financial statement. So if you do an income statement every month, you need a monthly entry for depreciation. You can make an estimate of the amount based on the total depreciation claimed in the past year. See Chapter 10 for more details.

Here's how depreciation works: Suppose you buy a capital asset like machinery, equipment or a truck. You either pay cash or finance the purchase. But because the effective and useful life of those assets is more than one year, you're not allowed to write them off as an expense in the year you bought them. They must be "recovered," as the tax law calls it, or depreciated over a period of years.

So each year, you record the appropriate amount of depreciation. In the asset section of your balance sheet, you set up a reserve account to reduce the book value of the asset. This is offset by the debit that goes to the depreciation expense account:

Date	Description	Debit	Credit
	Depreciation Expense	1,400.00	
	Reserve for Depreciation		1,400.00

Depreciation continues until the recovery period expires and the book value of assets is zero. This will take as little as three years or as long as 31.5 years (for commercial real estate).

4) Bad debts

On the accrual basis, you report income as it is earned even though you don't collect it until later. Unfortunately, experience teaches you that you'll never collect part of your total receivables. Because of this, you're allowed to set up a reduction of the receivable asset, called the reserve for bad debts.

Each month, you write off a pre-determined amount (or a percentage of total receivables.) The effect of this is to increase bad debt expenses, and to reduce the gross accounts receivable balance:

Date	Description	Debit	Credit
	Bad Debt	450.00	
	Reserve for Bad Debts		450.00

You can adjust the reserve periodically in one of two ways:

1) The fixed reserve amount is increased or decreased based on changing circumstances

2) A single account becomes a bad debt, and is written off in addition to the reserve amount.

The four non-cash adjustments described above are shown in Figure 5-4.

Managing the General Journal

Be certain that all journal entries you make are necessary. Don't abuse the adjustment journal, and always explain every entry you make.

There are ways to misuse the general journal. Avoid these:

- Entering journals to balance an account, rather than reconciling it. For example, you can't reconcile the bank account, so you make a journal entry to write off the

non–cash expense journals

DATE	EXPLANATION	✓	DEBIT	CREDIT
	-1-			
9-1	Insurance		193.00	
	Prepaid Assets			193.00
	to record 1/24th of two-year			
	insurance premium			
	-2-			
9-1	Amortization		78.45	
	Organizational Expenses			78.45
	to amortize 1/60th of total			
	start-up costs			
	-3-			
9-30	Depreciation Expense		1,400.00	
	Reserve for Depreciation			1,400.00
	-4-			
9-30	Bad Debts		450.00	
	Reserve for Bad Debts			450.00
	to record estimate of bad debts			

Figure 5-4

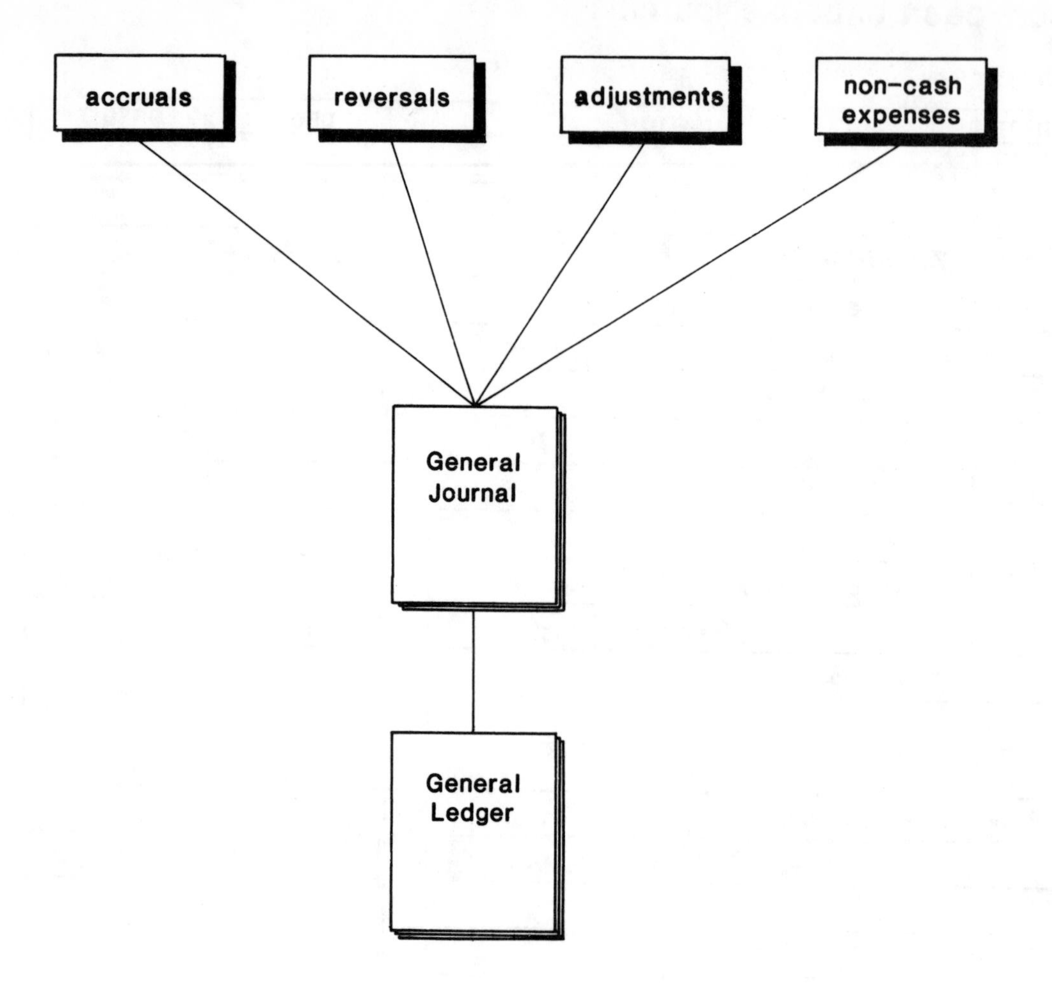

Figure 5-5

out-of-balance amount to miscellaneous expenses.

- Adjusting "to budget." If you have estimated an expense and the actual cost is higher, you should never prepare a journal to defer part of the actual expense. This is contrary to the purpose of budgeting, which is to identify problems.

- Preparing journals to create a specific level of profits. Don't manipulate the books so profit, cost or expense levels match your expectations. Never expect yourself to accurately predict the future. Estimates are just that. Using the general journal this way only delays facing a problem, and could make it worse. The purpose of the general journal is to make all entries to the general ledger as uniform as possible. It shows a trail of everything that goes into the final books of entry, from receipts and vouchers, through the journals, to the ledgers.

The four types of journal entries are illustrated in Figure 5-5.

In the next chapter, you will see how all your entries — from the receipts, payments and general journal — are brought together and posted to the general ledger.

Self Test

1. ***The purpose of the general journal is to:***

   a) record expenses that you pay out of pocket rather than through the checkbook or the petty cash fund.

   b) adjust for errors in coding.

   c) record non-cash expenses.

   d) all of the above.

2. ***An accrual journal for accounts payable:***

   a) should never be reversed, since it is the only time that liabilities will be recorded.

   b) must always be reversed, either by another journal or by coding of payments to accounts payable.

   c) must be reversed only if you later find an error.

   d) should not be made if it will only be reversed next month.

3. ***The purpose of an accrual journal is to:***

   a) book all income, costs and expenses as early in the year as possible.

   b) book all income, costs and expenses in the month earned or incurred, regardless of when received or paid.

   c) maximize profits you will report this month.

   d) none of the above.

4. ***The reversal journal is necessary when:***

   a) that is the most efficient way to reverse an accrual made last month.

   b) the original accrual turns out to contain errors.

   c) the financial statement shows a loss, and you need to report a profit.

   d) none of the above.

5. ***Adjustment journals are made when:***

   a) an accrual is left on the books by mistake.

   b) the budget is too low, and you need to defer some current expenses.

   c) you discover errors in coding or in the bank reconciliation.

   d) you need to accrue income.

6. ***Non-cash expense journals are made when:***

   a) you owe money, but do not have the funds to make payment.

   b) you need to update the status of accounts receivable, but the customer has not yet paid.

   c) you need to record expenses that won't be paid until next month.

   d) none of the above

Chapter 6

THE GENERAL LEDGER

At the end of each month, you post information from the journals to the general ledger, and then close the books. "Closing" the books means that you've recorded all the month's transactions, balanced the books and made all the adjusting entries. Then you're ready to prepare financial statements.

In the previous three chapters, you saw how three specialized forms of journals work. The income, or receipts journal, the payments journal, and the general journal are only posting sources for the general ledger. All entries made into each of the journals are transferred into the general ledger.

Rules for Managing the Ledger

The general ledger is the final book of entry. You'll manage it easily if you follow these rules:

1) Keep it as simple as possible

The more accounts or sub-accounts you have in the general ledger, the harder it is to balance. Put details in subsidiary ledgers (for accounts receivable, breakdowns by jobs, etc.) and summarize them into the general ledger.

You might have only two or three customers when you first open for business. In that case, you can have a separate accounts receivable account in the general ledger for each customer. But when you grow to 25 or 30 or more customers each year, your general ledger would become too bulky. The solution is to have only one general ledger page for accounts receivable. Use a subsidiary accounts receivable ledger for the details of individual customer accounts.

The general ledger isn't the place for a lot of details. It simply summarizes information needed for the financial statements. Organize the general ledger accounts in the same order as they appear in the statements. We'll talk about that under "The Posting Process" later in the chapter.

2) Post only once each month

Your posting has to be completely accurate if your books are going to balance. If you're using your journals properly, you can post the general ledger just once a month, and it won't take you very long. You transfer the balanced totals from each journal into the appropriate general ledger account. You'll make a relatively small number of entries each month.

The income and payments journals contain more detail than the general journal does. You can save a lot of time when you close the books each month by preparing ahead. For example, if you write many checks, you can add and cross-check the payments journal subtotals periodically. Then, at the end of the month, you have relatively few entries to check and balance. You can make the summarized payments journal entries in a few minutes.

3) Check balances forward

Before you start posting to your general ledger, be sure it's in balance to start with. Check all the balances forward. Then add all the debit balances and subtract all the credit balances. The total for the ledger should be zero. If it's not, find any mistakes before you transfer entries from the current month.

Also check the balances in every journal before you post entries. This is easy. In the payments and income journals, you just make sure that the detail columns add up to the same amount as the total column. And be sure that all debits equal all credits on general journal entries.

4) Consider the ledger the final word

The general ledger is a reference to your detailed subsidiary records. In your subsidiary account for accounts receivable, there's a card for each customer. The total of the month-end balances on those cards must always be the same as the balance in the accounts receivable account in your ledger — without exception. If these don't agree, find the error before you close the books.

All subsidiary records work the same way. They all have to balance with the figure reported in your general ledger. The general ledger is useful not only for the preparation of financial statements, but as a single controlling record — the final word on the balances in every other record you keep.

5) Reconcile all accounts

There are records that you keep outside of the general ledger. These are related to the asset and liability accounts. Asset records include lists of your capital assets (furnishings, equipment and property), your cash on hand (the bank reconciliation and petty cash fund), and accounts receivable.

Liability records include taxes payable (payroll, sales and use taxes), notes, and accounts payable. Make certain that all asset and liability accounts are reconciled, meaning you always know what is included in the current balance.

Each of the balances in these subsidiary records should always balance to the corresponding general ledger account.

If you can't explain the contents of any account, you don't have a complete set of books. Set up a procedure to reconcile every asset and liability account each month, and don't let any account get out of hand.

One of the bookkeeper's most important functions is to be sure the books are accurate. You must always be able to explain the contents of each account. So if an account doesn't balance or its entries haven't all been checked, it's out of hand. You need to reconcile that account and transfer any inappropriate entries to their correct place.

In Chapter 16, we'll look at the major asset and liability accounts and see how to keep them accurate.

Use the general ledger as a reconciliation control. When you balance those subsidiary records each month, be sure they agree with the reported totals in the general ledger.

The General Ledger Account

Each account in the general ledger has six distinct sections: the date, description, posting reference, debit, credit, and balance. It looks like this:

Date	Description	Ref.	Debit	Credit	Balance

The balance column consists of the balance forward, plus any debits posted for the month, and less any credits. You can set up the general ledger account in the same format as a journal.

The date column contains the actual posting date. If you post at the end of each month, this should always be the last day, "1-31" for example. However, if you make a journal entry on some other day, you should use that date.

Use the description column only when special notes are needed. For example, "balance forward" is usually written on the first line of balance sheet accounts, for balances carried over from the previous period. In some accounts, you might want to note a special circumstance, or a reminder to take some future action. An example would be a reference to a reversing entry.

The posting reference column gives you a trail back to the appropriate journal. Commonly used abbreviations are CR (Cash receipts), CD (Cash disbursements), and GJ (General journal). For completely accurate referencing, it's a good idea to refer to the journal entry number. For example, when you post the sixth journal entry for January, fill in the reference column with "J1-6."

You post the debits and credits in the next two columns. You enter the balance in the final column. The balance is the previous balance forward, plus all debits, less all credits. You should enter one balance for the end of each month, regardless of how many entries you posted.

Here's an example:

At the end of January, you post three entries to the Cash account. There's an entry from the receipts journal, one from the payments journal, and one from the general journal for an adjustment. At the end of the month, the cash account would look like this: You'll have a balance forward and a closing balance.

Cash					
Date	Description	Ref.	Debit	Credit	Balance
2-29	Balance forward				17,000
2-29		R2-1	1,900		
2-29		P2-1		2,500	
2-29	Bank charge	J2-3		34	16,366

In the general ledger, asset and expense accounts normally have a debit (positive) ending balance. Liabilities and income accounts normally have credit balances (negative amounts.) The formula is always the same for figuring the new balance forward.

debit-balance accounts:

\+ balance forward

\+ debits

– credits

= balance forward

credit-balance accounts:

– balance forward

\+ debits

– credits

= balance forward

A typical account, with four months posted, is shown in Figure 6-1.

Using Sub-accounts

Because general ledger accounts are summaries of more detailed records, there are a minimum number of posted entries. But sometimes you need a little more detailed information in the general ledger. At those times, sub-accounts can solve your problem.

A contractor I know prepares a monthly financial statement showing travel and entertainment as a single account. To be consistent, these expenses should be lumped together in the general ledger, too. However, only 80 percent of the entertainment expenses are deductible. In addition, he wants to track transportation expenses reimbursed to his foremen each month.

the general ledger account

CASH – CHECKING ACCOUNT

DATE		REF.	DEBIT	CREDIT	BALANCE
	Balance forward				4,316.45
1-31		CR	36,310.45		
✓		CD		37,403.62	
✓		J1-6	213.51		3,436.79
2-28		CR	34,910.06		
✓		CD		31,206.55	
✓		J2-4	306.00		7,446.30
3-31		CR	41,216.98		
✓		CD		40,994.13	
✓		J3-9		111.48	7,557.67
4-30		CR	52,403.66		
✓		CD		56,208.40	
✓		J4-5	32.15		3,785.08

Figure 6-1

He can collect that special information by using sub-accounts. This doesn't mean breaking the single account into three separate accounts. There's still only one "Travel and Entertainment" account in the general ledger. But the details of all posted debits and credits, and the balances forward, are broken down into categories.

Figure 6-2 shows how this works. On the right is a page from the ledger, showing the normal summarized entries each month. On the left is an additional page for breaking down every entry by sub-account.

Notice that every debit and credit, and the ending balance each month, reconciles to the penny with the sub-account postings. This is essential for good and complete management of the general ledger. Keep your sub-accounts in balance. It's a nuisance to have to go back and find your mistakes later.

There are times when sub-accounts should not be used, such as:

1) For multiple cash accounts. It's wiser to keep cash accounts separate. Your monthly reconciliation will be easier.

2) For accounts better controlled in subsidiary ledgers. These include accounts receivable, payroll, and accounts payable.

Subsidiary accounts are most practical when you have many similar records, as opposed to

the account with sub-accounts

Travel	Entertainment	Transportation	DATE	TRAVEL AND ENTERTAINMENT EXPENSE	REF.	DEBIT	CREDIT	BALANCE
–	45.50	107.90	1-31		CD	153.40		153.40
240.00	93.18	182.55	2-28		CD	515.73		
100.00	–	35.00	✓		J2-8	135.00		804.13
<100.00>	–	<35.00>	3-1		J3-1		135.00	
116.50	84.15	143.18	3-31		CD	343.83		1,012.96
–	103.45	115.00	4-30		CD	218.45		1,231.41
215.80	62.00	149.00	5-31		CD	426.80		1,658.21
–	–	118.50	6-30		CD	118.50		
–	14.00	162.00	✓		J6-7	176.00		1,952.71
–	<14.00>	<162.00>	7-1		J7-1		176.00	
395.00	118.06	284.15	7-31		CD	797.21		2,573.92
114.06	84.83	107.88	8-31		CD	306.77		2,880.69
1,081.36	591.17	1208.16		SUB-TOTAL				

Figure 6-2

breakdowns of an account into distinct types. Suppose you have a large number of charge customers. It wouldn't be practical to set up a sub-account under accounts receivable for every customer. Your general ledger would be hard to manage, since you could require dozens, or even hundreds of sub-accounts.

Don't get bogged down in breakdowns. Don't create sub-accounts when it only makes more work for you. Be sure there's a good reason, such as the need to monitor a budget, report differently for tax purposes, or prepare detailed financial statements.

Sub-accounts are especially helpful for payroll tax liabilities, which can be broken down into their four or five different types. In your financial statement, you only want to show one total for the liability. But you also need to oversee each state and federal debt to make sure you're always in balance.

The Posting Process

Once you set up your ledger and enter the beginning balances, you're ready to post from various journals. Remember that all entries must be offset by an equal entry or entries on the opposite side of the ledger. Once you've balanced journals, the posting itself is a fairly mechanical part of the job. All of the details have been isolated in journals, balanced, and summarized in closing totals.

The income journal is posted like this:

1) Total cash received is a debit to cash, and a credit to accounts receivable (for payments received on account) or credits to income (for cash sales).

2) Total charged sales are debits to accounts receivable, and credits to income.

The payments journal is posted as a credit to the cash account, and debits to the various accounts broken down in the journal.

Every entry in the general journal is posted to the appropriate account

Arrange your general ledger so each type of account is easy to find. The recommended order is the same as the accepted format for financial statements. Balance sheet accounts come first, followed by profit and loss accounts.

Figure 6-3 shows the sequence of the ledger.

The balance sheet accounts include all assets, liabilities and net worth. And the profit and loss accounts are for income, costs and expenses.

You can quickly find the appropriate account once you understand how each journal relates to the general ledger.

The income journal deals with two types of accounts: assets and income. You will post to the cash and accounts receivable accounts (both assets), and book current income from this journal.

The payments journal involves a credit to cash for the total amount paid out. The credit is offset by debits to various liability, cost and expense accounts.

The general journal can include entries to any account in the general ledger.

Figure 6-4 shows how each of the journals relates to the general ledger.

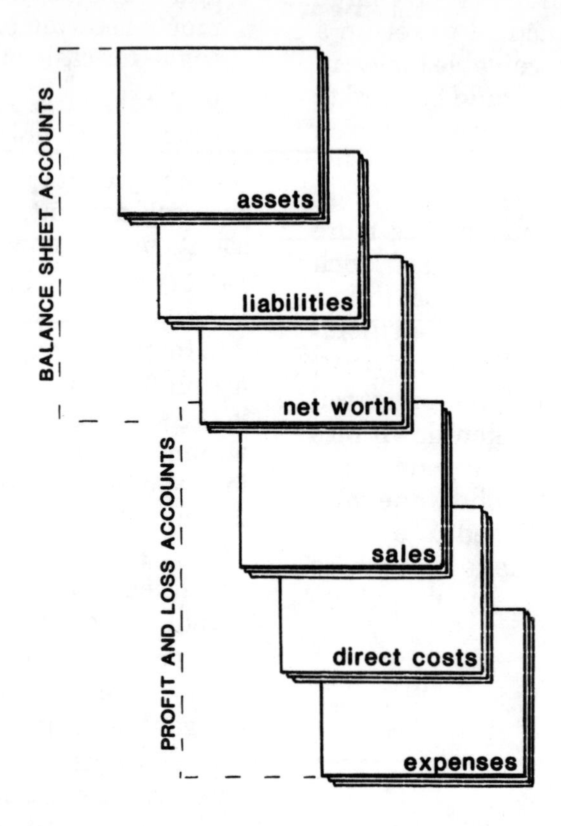

Figure 6-3

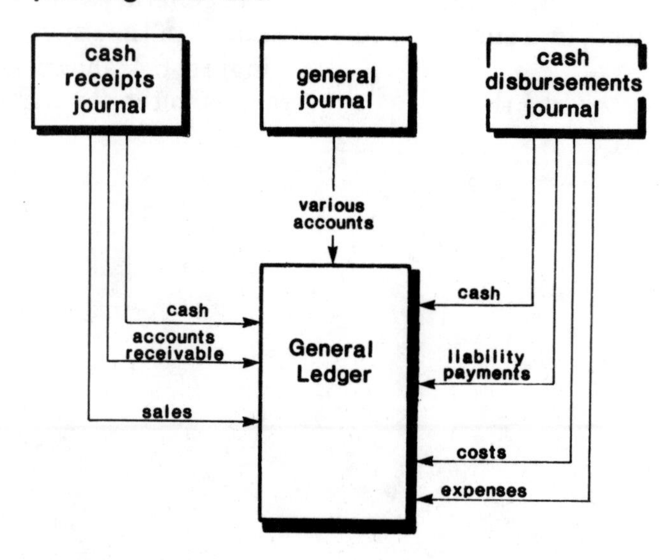

Figure 6-4

Self Test

As of January 1, the first month of the year, your books show balances forward in asset and liability accounts only. There is nothing posted to profit and loss accounts, because you're starting a new year. The following pages contain a sample general ledger for the new year. Post the entries from the income, payments and general journal that are summarized below.

Summary, income journal:		
	Debits	**Credits**
Cash	18,450.00	
Accounts receivable		18,450.00
Accounts receivable	22,485.10	
Income		22,485.10

Summary, payments journal:		
	Debits	**Credits**
Cash		19,942.78
Materials	9,216.84	
Direct Labor	4,532.00	
Operating Supplies	215.00	
Freight	85.40	
Rent	700.00	
Salaries	2,800.00	
Office Supplies	184.15	
Telephone	266.32	
Utilities	85.55	
Repairs	32.00	
Postage	44.00	
Subscriptions	15.40	
Travel	106.73	
Printing	84.77	
Advertising	102.00	
Professional Fees	400.00	
Licenses and Fees	15.00	
Insurance	106.00	
Note Payable	207.11	
Payroll Taxes Payable	744.51	

General journal:			
Date	**Description**	**Debit**	**Credit**
1-1	Depreciation Expense	1,400.00	
	Reserve for Depreciation		1,400.00
	–estimated depreciation–		
1-2	Insurance	38.15	
	Prepaid Assets		38.15
	–1/24th amortization–		
1-3	Payroll Tax Expense	704.16	
	Payroll Taxes Payable		704.16
	–adjust liability for January–		
1-4	Materials	1,101.65	
	Office Supplies	109.78	
	Telephone	211.50	
	Accounts Payable		1,422.93
	–liability at 1-31–		
1-5	Accounts Payable	1,143.83	
	Materials		945.70
	Office Supplies		55.18
	Operating Supplies		62.95
	Telephone		80.00
	–reverse 12-31 liability-		
1-6	Office Supplies	32.50	
	Bank Charges	8.00	
	Accounts Receivable		250.00
	Cash	209.50	
	–bank account adjustments–		

1) Post all of the entries in the three journals above, to general ledger accounts that follow.

Cash					
Date	**Description**	**Ref.**	**Debit**	**Credit**	**Balance**
	Balance Forward				1,996.80

Accounts Receivable					
Date	**Description**	**Ref.**	**Debit**	**Credit**	**Balance**
	Balance Forward				22,116.06

Inventory					
Date	**Description**	**Ref.**	**Debit**	**Credit**	**Balance**
	Balance Forward				14,300.00

Fixed Assets					
Date	**Description**	**Ref.**	**Debit**	**Credit**	**Balance**
	Balance Forward				62,814.00

Reserve For Depreciation

Date	Description	Ref.	Debit	Credit	Balance
	Balance Forward				(16,811.00)

Prepaid Assets

Date	Description	Ref.	Debit	Credit	Balance
	Balance Forward				648.55

Accounts Payable

Date	Description	Ref.	Debit	Credit	Balance
	Balance Forward				(1,143.83)

Payroll Taxes Payable

Date	Description	Ref.	Debit	Credit	Balance
	Balance Forward				(744.51)

Notes Payable

Date	Description	Ref.	Debit	Credit	Balance
	Balance Forward				(956.16)

Net Worth					
Date	Description	Ref.	Debit	Credit	Balance
	Balance Forward				(82,219.91)

Income					
Date	Description	Ref.	Debit	Credit	Balance

Materials					
Date	Description	Ref.	Debit	Credit	Balance

Direct Labor					
Date	Description	Ref.	Debit	Credit	Balance

Operating Supplies					
Date	Description	Ref.	Debit	Credit	Balance

Freight					
Date	**Description**	**Ref.**	**Debit**	**Credit**	**Balance**

Rent					
Date	**Description**	**Ref.**	**Debit**	**Credit**	**Balance**

Salaries					
Datd	**Description**	**Ref.**	**Debit**	**Credit**	**Balance**

Payroll Taxes					
Date	**Description**	**Ref.**	**Debit**	**Credit**	**Balance**

Office Supplies					
Date	**Description**	**Ref.**	**Debit**	**Credit**	**Balance**

Bank Charges					
Date	**Description**	**Ref.**	**Debit**	**Credit**	**Balance**

Telephone					
Date	**Description**	**Ref.**	**Debit**	**Credit**	**Balance**

Utilities					
Date	**Description**	**Ref.**	**Debit**	**Credit**	**Balance**

Repairs					
Date	**Description**	**Ref.**	**Debit**	**Credit**	**Balance**

Postage					
Date	**Description**	**Ref.**	**Debit**	**Credit**	**Balance**

Subscriptions					
Date	**Description**	**Ref.**	**Debit**	**Credit**	**Balance**

Travel					
Date	**Description**	**Ref.**	**Debit**	**Credit**	**Balance**

Printing					
Date	Description	Ref.	Debit	Credit	Balance

Advertising					
Date	Description	Ref.	Debit	Credit	Balance

Professional Fees					
Date	Description	Ref.	Debit	Credit	Balance

Licenses and Fees					
Date	Description	Ref.	Debit	Credit	Balance

Insurance					
Date	Description	Ref.	Debit	Credit	Balance

Depreciation					
Date	Description	Ref.	Debit	Credit	Balance

2) Prove the accuracy of your posting by listing the balances of each account on the schedule below. Be sure the totals of debits and credits are the same, to prove you are in balance.

	Debit	Credit
Cash		
Accounts Receivable		
Inventory		
Fixed Assets		
Reserve for Depreciation		
Prepaid Assets		
Accounts Payable		
Payroll Taxes Payable		
Notes Payable		
Net Worth		
Income		
Materials		
Direct Labor		
Operating Supplies		
Freight		
Rent		
Salaries		
Payroll Taxes		
Office Supplies		
Bank Charges		
Telephone		
Utilities		
Repairs		
Postage		
Subscriptions		
Travel		
Printing		
Advertising		
Professional Fees		
Licenses and Fees		
Insurance		
Depreciation		
TOTALS		

CHAPTER 7

BALANCING AND CLOSING THE BOOKS

When you're new to bookkeeping, you're wise to take your time and pay a lot of attention to accuracy. At first, if your books don't balance, you won't have the experience to know the most likely places to find mistakes. That will come with time. Everyone makes mistakes sometimes. The trick is to make as few as possible, and correct them easily when you do.

Common Bookkeeping Errors

Remember that the purpose of bookkeeping rules is to make sure that all transactions are posted accurately. If you make an error, it shows up at once. There are techniques for finding and correcting mistakes.

Here are some common posting errors:

1) Posting from a journal that doesn't balance. When you do that, the ledger won't balance either.

2) Posting a debit as a credit, or vice versa. In this case, your general ledger will be out of balance by twice the amount of the posted transaction.

3) Transposition. An example would be if you posted $812.15 as $821.15. If you're out of balance by any number whose digits add up to 9 or a multiple of 9, you probably have a transposition error. The difference between 21 and 12 in the example above is 9. The difference between any transposed numbers will also be divisible by 9.

4) Math errors in the account. You might enter a number correctly, but then add or subtract the account record wrong.

5) Leaving out an entry. As you post each entry, you should mark the journal's reference column with a checkmark or the account code. In the receipts or payments journals, you'd put the reference mark at the bottom of the account column.

6) Posting to the general ledger when its balance forward is wrong. In this case, you won't balance, no matter how accurately you post.

Remember that a correctly posted general ledger always adds up to zero. That is, the sum of all the debits equals the sum of all the credits. If everything has been posted correctly and there aren't any addition errors, a zero balance will prove there aren't any mistakes, or omitted or duplicate entries.

With that in mind, follow this procedure whenever you post:

1) Check the general ledger's balance forward. This is especially important if you're not familiar with the general ledger, or if someone else has been posting it before you. Total the ending balances (balances forward) before you start posting the current entries from the journals.

It's easier to keep the ledger in balance if you only post once a month. That limits the chance for mistakes when you make only one entry per month from each of the journals.

When you're the only one in control of the general ledger from one month to the next, you don't need to double-check the balance forward if the ledger balanced at the end of the last posting period.

2) Check the journals before you post. Be sure the total debits equal the total credits. Even check journal entries with only one debit and one credit. A carelessly posted journal could contain the following entry:

Date	Explanation	Debit	Credit
	Insurance Expense	316.45	
	Prepaid Expenses		314.65

A quick glance at this account isn't enough to catch the subtle transposition. But a careful review of your entries *before* posting helps avoid errors.

3) Post carefully. You'll make fewer mistakes if you do all the posting at one time. This isn't always practical, especially if you have other duties. You'll be interrupted by phone calls and visitors and be distracted by other duties. All this can lead to posting errors.

Try to find a block of quiet time to do your posting from the journals to the ledger. If that's not possible, you have to take great care to be accurate.

4) Check the general ledger's ending balances. Add or subtract the ending balances in each account as appropriate. Remember that a debit-balance account is a plus and a credit-balance account is a minus. A credit balance forward, following by an additional credit this month, creates a larger credit balance. The income account is an example of this.

Income					
Date	**Explanation**	**Ref.**	**Debit**	**Credit**	**Balance**
5-1	Balance Forward				(15,000)
5-29	Charge Sales	J1		20,000	(35,000)

If your general ledger balances the first time, your job is done. All entries were correctly posted and brought forward. However, if the ledger doesn't balance, stop now and correct any mistakes before going on.

Don't overlook the obvious here. The ledger may actually be in balance, but you could have made a mistake when adding the totals. You can save yourself a lot of time by running this tape again. At least you'll verify the exact amount you're out of balance, if in fact you are.

If you really don't balance, go on with the following steps:

5) Check all posted entries. Trace each entry back to its journal. The most common errors occur when you copy numbers from the journal to the ledger.

If you have a large number of posted transactions, here's how you can check two things at the same time and speed up the checking process. Enter each current general ledger transaction into your adding machine, being sure to "plus" the debits and "minus" the credits. If you do this correctly, you should end up with a zero balance.

Even when your journals are in balance, you can still make a mistake when you transfer the posted numbers to the ledger. So you run the tape to verify the journal totals (see step 1 above) *and* your posting. You compare the entries on your balanced adding machine tape to the amounts actually posted in the ledger.

If your tape total isn't zero, you have an error, either in running the tape or in one of your entries. Find and correct the error, and then proceed.

With a verified tape, compare each entry to the amounts posted in the general ledger. You'll find any misposted entries this way.

Even when you have a large number of accounts in your ledger and a lot of transactions, you can still prove current entries with the tape. You don't know which account each tape entry applies to. But you can relate the tape to all currently posted amounts. Suppose you have posted three entries in your cash account. Find those amounts on your adding machine tape and cross them off the tape. You'll find any tape or posting error by the process of elimination.

6) Check the math in the general ledger. Chances are, your error was incorrect addition or subtraction, a transposition, or turning an entry around (adding a credit or subtracting a debit).

The best way to find errors is to go through these steps in order. You start by checking balances forward and current debit and credit totals, then examine your calculations. If you're still out of balance, repeat the steps. You've overlooked something. Remember to check:

- general ledger balances forward
- balance of entries on the journals
- amounts posted versus journal amounts
- math in the ledger accounts

An Example of Posting Errors

Look at the journal and four accounts in Figure 7-1. This is a simplified version of the posting process. In reality, you'll probably post many general journal entries, as well as totals from receipt and payment journals. But to identify typical errors, this is a worthwhile exercise.

Find these five mistakes:

1) First, there's a math error in the journal. The total of the debits in that entry is really $4,618.25, so the journal itself is out of balance by $100. To correct this, don't automatically assume that the credit is wrong. The error could be in one of the debits. You have to go back to the source documents for the journal to find the actual error.

2) The cash account is added incorrectly. As the account is now posted, the correct new balance is $1,699.75, but remember that this could change once the error in the journal has been corrected.

3) There is a transposition error in the entry to the materials account. $4,316.85 was posted as $4,361.85.

If that had been the only error, you would have been out of balance by $45. If you remember the rule about transpositions, you'd suspect one in this case, since 4 + 5 = 9. And since 45 divided by 9 = 5, you'd know to look for an entry in the first two positions left of the decimal point where the two digits have a difference of 5, in this case 6 and 1. Entries with 27 and 72, 83 and 38, 94 and 49, and so on, would also be suspect.

4) The office supplies account contains an addition error of one dollar. The correct total is $1,351.75.

5) The entry in the telephone account was posted as a credit rather than as a debit. The ending balance is off by $200, twice the amount of the error.

This is a small-scale example of the kinds of errors you might find in your own general ledger. You can find the math errors easily by double-checking the accounts themselves. The other errors become obvious when you run two adding machine tapes, one of the journal entries and one

posting errors

JOURNAL		
Account	Debit	Credit
Materials	4,316.85	
Office Supplies	201.40	
Telephone	100.00	
Cash		4,518.25

CASH		
forward	6,218.00	
		4518.25
balance	1,697.75	

MATERIALS		
forward	12,954.16	
	4,361.85	
balance	17,316.01	

OFFICE SUPPLIES		
forward	1,150.35	
	201.40	
balance	1,350.75	

TELEPHONE		
forward	3,812.19	
		100.00
balance	3,712.19	

Figure 7-1

looking for errors

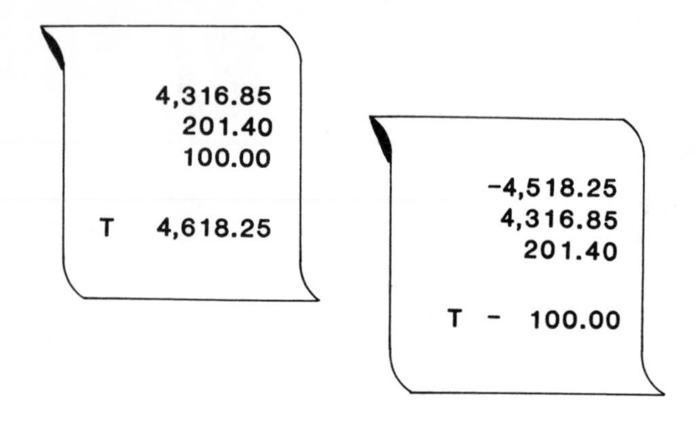

Figure 7-2

of the amounts posted to the ledger. Figure 7-2 shows the two tapes.

The tape on the left is a summary of the journal. By adding the debits, you immediately discover that the credit on the journal is not correct.

The tape on the right shows the entries as they were posted. Only the transposition error in the materials account has already been corrected, so the totals reflect the amounts posted from the journal (even though that journal still contains an error). The balance of minus $100 would be balanced out by a debit of $100; but the amount posted to the telephone account is a credit. With this tape and a quick check of the journal and math in the actual accounts, all of the errors will have been found and corrected.

The Trial Balance

Keep your journals up to date throughout the month. This will lighten your load at month-end so mistakes won't be as likely. Then, at the end of the month, balance and post all of the journals at the same time.

Once you've balanced the general ledger, it's time to prepare financial statements. The first step in making up the balance sheet and income (or profit and loss) statement is to strike a trial balance.

This is nothing more than a listing of all the accounts in the general ledger. Its purpose is to prove the balance and to identify net profit at the end of the period.

You prepare the trial balance in two sections: balance sheet accounts and profit and loss accounts, as shown in Figure 7-3. You'll notice there's a space for a net subtotal at the end of each section.

The calculated subtotal of all balance sheet accounts must equal the subtotal of all income (or profit and loss) accounts.

Balance sheet accounts include assets, liabilities and net worth.

Income statement accounts are income (sales), direct costs and expenses.

Remember that if your general ledger is in balance, the total of *all* accounts will be zero. But if you divide the ledger into these two sections, the subtotal of each section will *not* be zero.

That's because in the double-entry system, debits to balance sheet accounts are often offset by credits to income statement accounts, and vice versa. Let's look at the cash account as an example. Cash receipts are a debit to cash (a balance sheet account — an asset) and a credit to income (an income statement account).

Cash payments, on the other hand, are credits to cash (they reduce the asset which is a balance sheet account) and debits to various cost and expense accounts (income statement accounts).

For that reason, the difference between debits and credits on the balance sheet section of the trial balance will be the same as the difference between those on the income statement section. That difference will appear on opposite sides of the ledger.

This isolated net number is the amount of profit (or loss) in the general ledger. If the business makes a profit, the subtotals will appear as shown in Figure 7-3. That is, debits will be greater than credits in the balance sheet section, while credits will be greater in the income statement section, by the same amount. If there's a loss, the situation will be just the opposite.

The trial balance is a useful tool for preparing financial statements from proven totals, but it's not essential. An experienced bookkeeper can often prepare the financial statements directly from the general ledger.

The trial balance is more important when the bookkeeper posts the ledger for an accountant to prepare the statements. The accountant doesn't

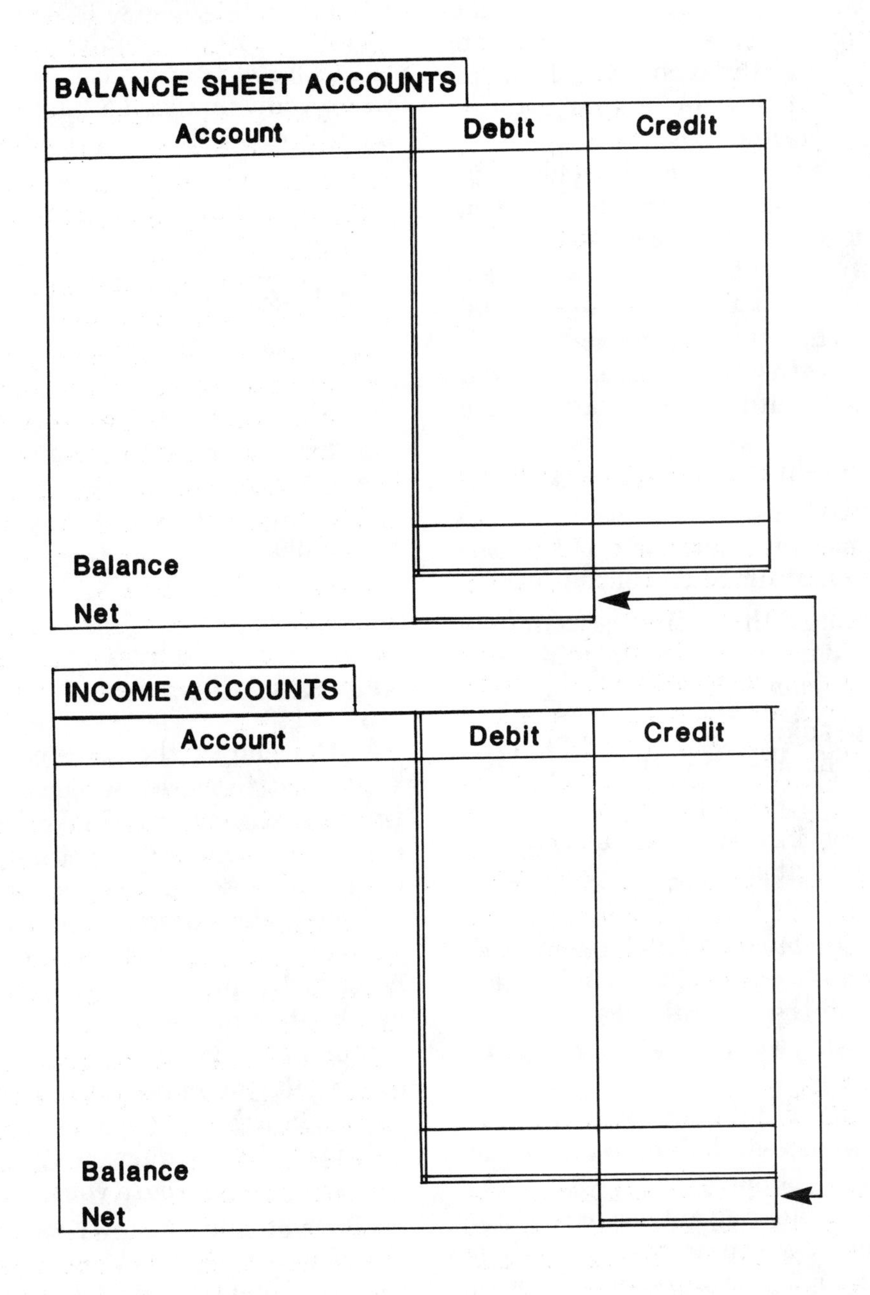

Figure 7-3

need to verify the balance by going through the entire general ledger. It's all summarized on the trial balance.

You use a trial balance to verify totals and to prove your posting is correct before you proceed to the financial statements. You can also post adjusting entries on the same page as the trial balance before you prepare the statements. So a trial balance serves two purposes. It's a verification tool and a worksheet.

Adjusting Entries

When you're finished posting the general ledger, you're almost ready to prepare a financial statement. But if you remember our discussion of accrual accounting, you know you have to make certain adjustments in order for those statements to be accurate.

Some adjustments aren't normally made until year end. But if you want an accurate interim statement, you need to make some adjustments before you prepare the statement. This can present a problem. If you make adjusting journal entries each month, your general ledger will become cluttered with estimated entries you don't want in there until 12 full months have passed.

Consider the following:

1) You want to show depreciation expense on your monthly statement, but that entry isn't posted until the end of the year.

2) You must adjust the liability account for the estimated interest expense paid year to date on a loan.

3) You need to adjust payroll taxes to show the actual liability as of the end of the month.

4) The inventory in your warehouse now is different from that shown in the general ledger.

These adjustments must be made before preparing a financial statement, but you'd rather not include them in the general ledger. In this case, prepare a closing worksheet like the one in Figure 7-4.

You list each account in the first column, with the closing debit and credit balances under the two "general ledger" columns.

Next, you put the adjusting entries in the "adjustments" column. It's a smart idea to explain these entries with another worksheet-form of journal entries. This worksheet should look just like the general journal, although entries are made only on the worksheet, and not in the general ledger itself.

If you only have a few adjusting entries, you don't have to prepare a journal-form explanation. You can simply explain adjustments with a footnote to the worksheet. Use a small letter in parentheses to reference each adjusting entry to its footnote.

Next, combine the two sections to arrive at the "final balance." This is the adjusted total you'll use on the financial statements.

Suppose you estimate your depreciation expense at $718 per month. At the end of May, you want your financial statement to reflect five months' depreciation. The adjusting entry would be:

Date	Description	Debit	Credit
5-31	Depreciation Expense	3,590.00	
	Reserve for Depreciation		3,590.00

On the worksheet, the Reserve for Depreciation account would be adjusted (increased) by the amount of the credit; and the Depreciation Expense account would show an adjustment of $3,590.00.

Year-end Adjustments

At the end of the year you'd use the same worksheet to close the books and prepare the financial statements. The difference in this case is that adjustments are actually put through the books and entered into the general ledger.

You'd make the same kinds of adjustments at year end as those for the interim statements described above. But now you have the actual yearly amounts, and not just estimates. Before you prepare the final balance sheet and income statement, you make the final adjusting journal entries like those shown in Figure 7-5.

You post these adjustments to the general ledger *before* you prepare your trial balance.

The depreciation entry records depreciation expense for the year, and also increases the reserve. The Reserve for Depreciation is a "negative" asset account, as shown in Chapter 2. The total of all fixed assets is reduced by the amount of the reserve.

Interest expense is reported annually in this example because you don't usually have an accurate monthly breakdown. Each month, the total note payment was applied against the liability account. At the end of each year, the ad-

closing worksheet

Date ____________

ACCOUNT	GENERAL LEDGER		ADJUSTMENTS		FINAL BALANCE	
	Debit	Credit	Debit	Credit	Debit	Credit

Figure 7-4

adjusting entries

DATE	EXPLANATION	DEBIT	CREDIT
	-1-		
12-31	Depreciation Expense	8,618.00	
	Reserve for depreciation		8,618.00
	to record annual depreciation		
	-2-		
12-31	Interest Expense	612.19	
	Notes Payable		612.19
	To adjust liability for annual interest		
	-3-		
12-31	Payroll Tax Expense	1,963.85	
	Payroll Taxes Payable		1,963.85
	to adjust year-end tax liability		
	-4-		
12-31	Cost of Goods Sold	16,500.00	
	Inventory		16,500.00
	to reverse beginning inventory		
	-5-		
12-31	Inventory	19,100.00	
	Cost of Goods Sold		19,100.00
	to record ending inventory		

Figure 7-5

the final journal

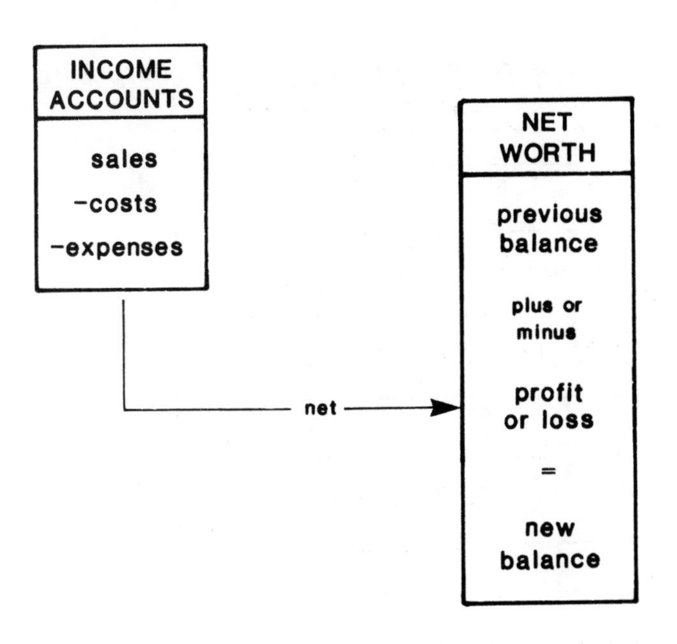

Figure 7-6

justing journal reports the amount of interest expense and restores the liability to its actual level.

You adjust payroll tax expense at the end of the year to show the actual amount paid and to adjust the liability level.

Inventory Adjustments

The last two entries on Figure 7-5 show how you reverse the old inventory level and record the new one. At the end of each year, you'd physically count your inventory and make this adjustment. Then you carry inventory as a current asset during the year, using the worksheet to make monthly adjustments. You'd change the inventory value in the general ledger only at the end of the year.

In the next chapter we'll discuss the preparation of the income statement in detail. You'll see that the income statement includes a "cost of goods" section that shows expenses for materials, direct labor, and other costs. It also involves adding the beginning inventory, and subtracting the ending inventory. The net difference between those values accounts for material purchases when the amount left on hand changes from one year to the next. The asset account "inventory" is changed to reflect the actual year-end inventory value.

During the year itself, the previous year's closing inventory is left on the books, even though monthly adjustments can be made to reflect changing inventory levels.

Transfer of Income Statement Accounts

You make a final journal entry at the end of each year to actually close out the profit and loss accounts. You carry forward the general ledger balances in the asset, liability and net worth accounts to the following year. But each year, the income accounts — sales, costs and expenses — start from a zero balance.

The final journal reverses the balances of each income account to zero, and offsets the net total with a single entry to "profit and loss," which is part of the net worth shown on the balance sheet. A profit increases net worth, while a loss decreases it.

For a year-end profit of $19,817, your final journal entry would look this way:

Date	Explanation	Debit	Credit
12-31	Income	493,651	
	Cost account		298,413
	Expense accounts		175,421
	Net worth		19,817
	– to record year-end profit –		

The final credit to net worth is the net profit for the year. It increases the total of the net worth account. When you've posted this entry, the general ledger will still balance, even though the income accounts have been closed to zero for the year.

Now you open your books for the new year with no sales, costs or expenses carried forward, only a balanced general ledger with assets, liabilities and an adjusted figure for net worth. This is illustrated in Figure 7-6.

In the next chapter, we'll see what all this has been leading to, the three types of financial statements, how to prepare them, and what they mean.

Self Test

1. Refer back to Figure 7-1, and fix the journal and each account so that it's posted correctly. Use the forms below for this:

Journal		
Account	Debit	Credit

Cash		
	Debit	Credit

Materials		
	Debit	Credit

Office Supplies		
	Debit	Credit

Telephone		
	Debit	Credit

2. If your general ledger is out of balance, you should check:

a) the balances forward in the general ledger, in case there are errors there.

b) amounts posted and the addition and subtraction in each account.

c) the accuracy of entries on journals, in case they are out of balance.

d) all of the above.

3. *Your general ledger is out of balance by $1,440.00. You should first check for:*

a) errors in balances forward in the general ledger.

b) errors on your adding machine tape.

c) transposition errors in posting or adding of accounts.

d) math errors in journal entries.

4. *The purpose of the trial balance is to:*

a) verify and prove the balance in each account of the general ledger.

b) find and correct posting entries before statements are prepared.

c) avoid the extra work of preparing financial statements every month.

d) all of the above.

5. *The closing worksheet is used to:*

a) verify account balances.

b) make entries to the general ledger.

c) make changes to totals without actually entering those changes in the general ledger.

d) all of the above.

6. *The final entry is:*

a) the last entry made each month in the general ledger.

b) closing of all balance sheet accounts to net worth.

c) closing of all income accounts to net worth.

d) zeroing out of every account in the general ledger.

Chapter 8

FINANCIAL STATEMENTS

Let's review for a moment the steps that lead to the preparation of the financial statements. There are three types of journal: the receipts journal, the payments journal, and the general journal. Each has its own purpose.

The receipts journal summarizes earnings and money received during the month. It's supported by details in a subsidiary accounts receivable system, if you use one.

The payments (disbursements) journal summarizes checks written during the month.

The general journal holds corrections, adjustments, non-cash expenses, and accruals.

The journals are posted to the general ledger, preferably once each month. The ledger is the final book of entry, with one page for each account. Balance sheet accounts — assets, liabilities, and net worth — are perpetual. Balances are carried forward from one year to the next. Income accounts are closed out to zero at the end of each year.

The trial balance proves the accuracy and completeness of the posted general ledger.

The worksheet is used for any adjustments needed for interim statements, but not entered into the general ledger. It's also used at year end to prepare closing statements.

Finally, financial statements report the results of operations and status of accounts as of a specific moment. They're usually prepared at the end of a month or business year. This entire chain is summarized in Figure 8-1.

Now let's go on to the preparation of the financial statements themselves. These are the final summaries of all the entries made during a reporting period. The entries you make in the books support and prove the conclusions listed on the statements. If there are questions about those conclusions, you'll be able to trace everything back from the ledger, through the journals, all the way to source documents (vouchers, invoices and receipts).

There is a distinction between the bookkeeper's job and the accountant's that becomes most obvious where financial statements are concerned. In some companies, the book-the posting and closing chain

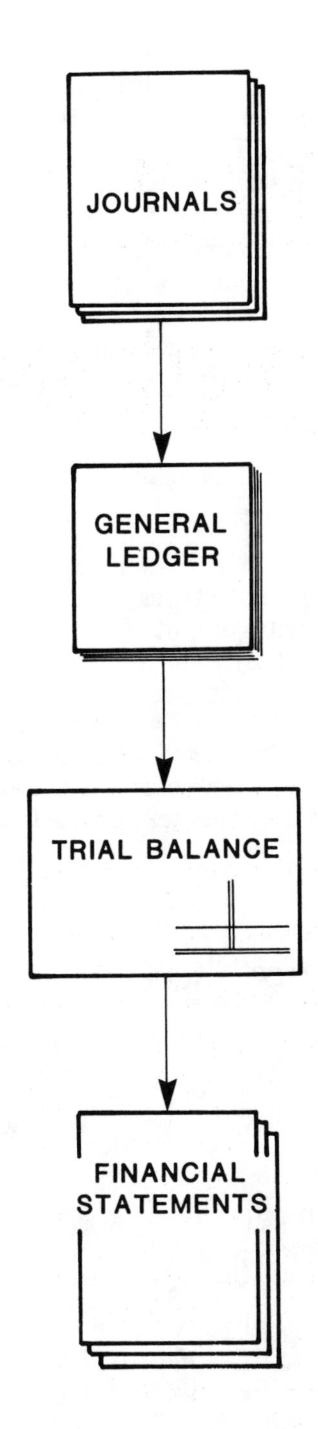

Figure 8-1

keeper goes only as far as posting the general ledger. Then the accountant prepares the statements and analyzes them for the business owner.

In other companies, the bookkeeper completes the statements. Then the owner or the accountant interprets the reports for financial and tax planning strategies.

Financial Statement Standards

There are three types of financial statements:

1) The Balance Sheet is a summary of all assets, liabilities and net worth, as of the statement date.

2) The Income Statement is also called the Profit and Loss Statement. This is a report of income, costs, expenses, and net profit or loss for the period ending on the statement date.

3) The Cash Flow Statement is also known as the Sources and Applications of Funds Statement. It shows how cash was accumulated and used during the time before the statement date. It also shows how working capital (the difference between current assets and current liabilities) has changed since the last reporting period.

All the statements must cover the same period, with the same beginning and ending dates.

Be consistent with the format for your statement headings. The accepted layout has the company name on the first line, the statement title on the second line, and the date or period covered on the third, with all of them centered at the top of the page.

ABC Construction Company
Balance Sheet
December 31, 1988

ABC Construction Company
Income Statement
January 1, 1988 to December 31, 1988

ABC Construction Company
Cash Flow Statement
January 1, 1988 to December 31, 1988

For income and cash flow statements, you can use a different format for the date:

For the year ending December 31, 1988

For comparative statements showing results for two years, for example, the date line should be specific. . .

For the 12 months ending 12/31/88 and 12/31/87

. . .or you can leave the third line off and include the date information above the information columns:

For the 12 months ending:
12/31/88 12/31/87

Underline all columns of numbers preceding subtotals, and put a double line under final totals (total assets and the sum of liabilities plus net worth, for example; or net profit on the income statement).

When you have two columns of numbers to report, indent the first subtotal column approximately the length of the numbers, so that one group doesn't interfere with the other. For example:

Incorrect:

Current Assets:	
Cash	$ 1,456.15
Accounts Receivable	37,403.01
Inventory	16,000.00
Total Current Assets	$54,859.16

Correct:

Current Assets:		
Cash	$ 1,456.15	
Accounts Receivable	37,403.01	
Inventory	16,000.00	
Total Current Assets		$54,859.16

Sources for Information

You find all the information you use for a financial statement in your general ledger. As I've said in previous chapters, it's absolutely necessary to begin with an accurate and completely posted general ledger. Be sure that the books are in balance before you begin. Otherwise, the financial statements won't be accurate or useful.

The balance sheet and income statement are the two primary statements. The balance sheet format uses the asset, liability and net worth accounts, including a final adjustment for profits during the year. The income statement summarizes sales, costs and expenses, and shows the net profit or loss for the year. Net income, adjusted by changes in asset and liability accounts, makes up the cash flow statement. Figure 8-2 illustrates this.

You arrange the general ledger in the same order that information is transferred to the financial statements. Here's why:

1) It makes it easier to prepare and analyze the statements.

2) The order is logical, so you can find accounts quickly. Once you become familiar with the arrangement of the ledger, it will be easy to post by type of account (assets or expenses, for example).

3) It's an accepted standard. If you're audited, or if a new employee takes over the books, the arrangement of the ledger will be familiar.

The Balance Sheet

The first of the three financial statements reports on the current balances in each account in the asset, liability and net worth sections of the ledger.

To review; the assets are property the company owns. Liabilities are the debts owed by the company. And the net worth is the amount of the owner's equity in the company.

Net worth is always the difference between assets and liabilities. On the balance sheet, assets are shown first, and then the combination of liabilities and net worth.

Assets = Liabilities + Net Worth

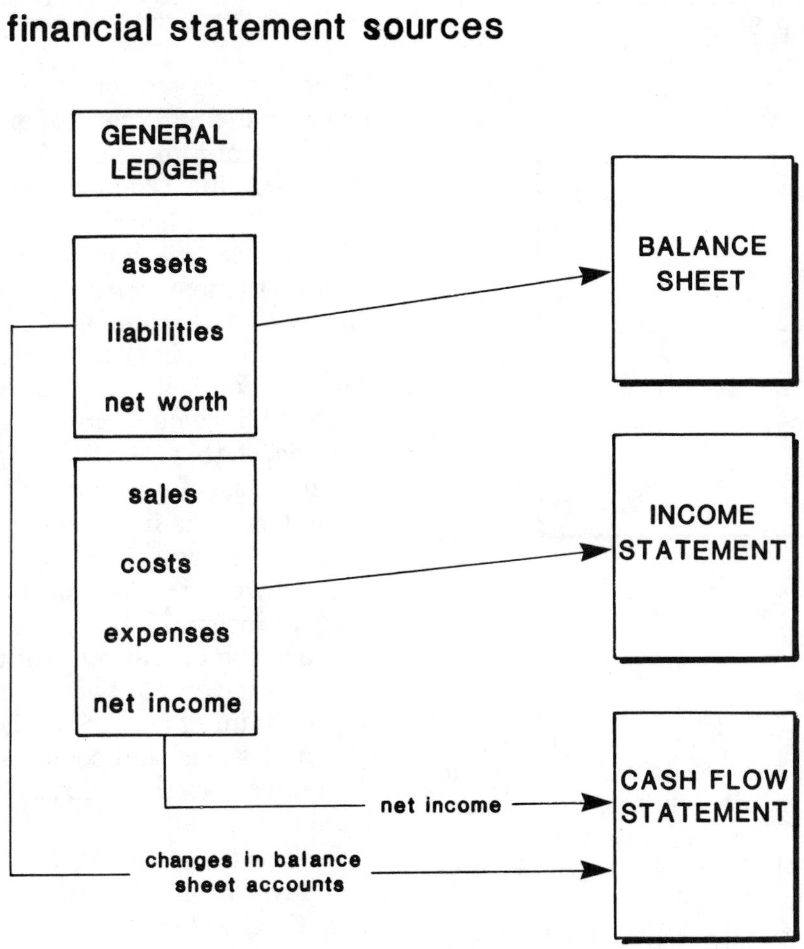

financial statement sources

GENERAL LEDGER

assets
liabilities
net worth

sales
costs
expenses
net income

BALANCE SHEET

INCOME STATEMENT

CASH FLOW STATEMENT

net income

changes in balance sheet accounts

Figure 8-2

The reason these balance is because profit earned during a period is added to net worth, or a loss is deducted from it.

Net Worth beginning balance + Profits = Net Worth ending balance

You'll recall that the trial balance will always be zero as long as there aren't any mistakes in the general ledger. The trial balance is split into two sections that correspond to the balance sheet and income statement. So the net total of all asset, liability, and net worth accounts is also the same net total as the one for the sum of sales, costs and expenses.

The balance sheet is broken down into sub-sections for the different types of assets and liabilities:

Assets

Current assets are cash, or assets that are convertible to cash within one year or less. This includes accounts receivable and inventory.

Long-term or "fixed" assets are the cost of all trucks, equipment, machinery and furniture. These are "capital assets" with a useful life beyond one year. The total of long-term assets is reduced by the accumulated depreciation claimed over the time those assets are owned.

Other assets include startup costs, goodwill, and other intangible property of the business. If you've recorded prepaid assets that are amortized over time, you'd show them here. Prepaid expenses are also assets.

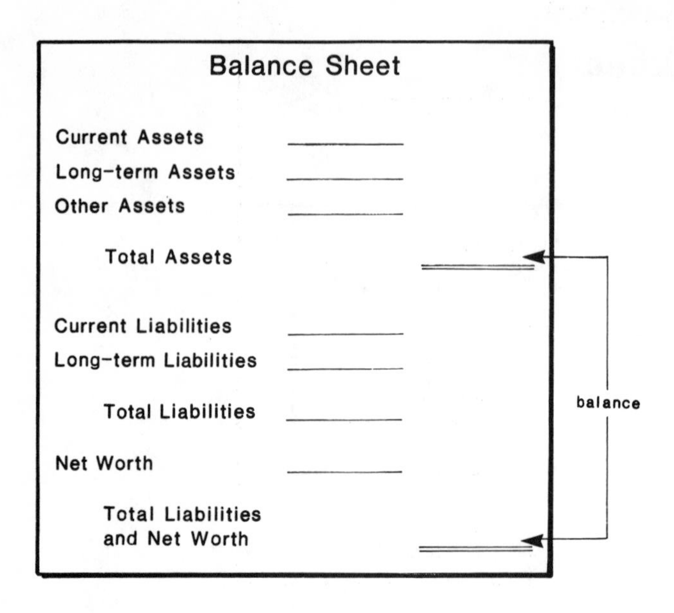

Figure 8-3

Liabilities

Current liabilities include debts that are due and payable in 12 months or less. You should include 12 months' worth of payments on all notes in current liabilities, with any overage included in the next category.

Long-term liabilities are debts owed beyond the coming 12 months. This would be any note balances that won't be retired during the coming year.

Deferred credits include income received before it's earned, such as down payments on contracts or early progress payments. While not truly a liability, you'd still report deferred income here.

Net Worth

This is any money the owner has contributed to the business, plus each year's profits, and minus any losses. In a sole proprietorship or partnership, net worth is reduced by the amount of money drawn out of the business by the owner.

Figure 8-3 shows the accepted format for a balance sheet. Note the reference to the balance: the total of assets must always equal liabilities plus net worth.

The Income Statement

The income statement reports sales, cost of goods sold, and expenses, and then summarizes the net profit for a specific period. At the end of the year, a full 12 months of activity is reported.

It's a good idea to report income, costs and expenses on a comparative basis. The income statement means more when you can see figures for two years side by side. If you prepare a statement during the year, compare it to the same period last year. For example, if you prepare an income statement for the first five months of this year, the figures for last year should be as of the same date.

The format is like the one in Figure 8-4. The first line shows sales (income), followed by a breakdown for cost of goods sold. The difference is gross income. Then you subtract expenses to produce the operating profit or loss. Next, you subtract the amount of federal income tax owed. The remaining amount is the net profit or loss.

Cost of goods sold consists of all direct costs, including materials, direct labor, and other related items.

You must also adjust for the change in inventory level by adding the beginning inventory, and then subtracting the ending inventory. Here's why:

Suppose you spend $80,000 on materials during the year. But the value of your inventory is $5,000 more than at the beginning of the year. Your true material cost is only $75,000, because $5,000 of your year's purchases are still in stock.

On the other hand, if inventory level at the end of the year is lower, then material costs are more than the amount spent. You've used up some of the stock you had on hand at the beginning of the year.

The formula looks like this:

Cost of goods sold:

  Beginning inventory
+ Materials purchased
+ Direct labor
+ Other direct costs
    Total
– Ending inventory
=   Cost of goods sold

Income Statement	
Sales	________
Less: Cost of Goods Sold	________
Gross Profit	________
Less: General Expenses	________
Net Operating Profit	________
Less: Federal Income Tax	________
Net Profit	========

Figure 8-4

You subtract the adjusted total of all costs from sales to get gross profit.

Sales - Cost of goods sold = Gross profit

Now you deduct all expenses from the gross profit. You can either list all expenses by account, or summarize them and attach a supplementary schedule. You can also classify expenses as either fixed or variable. Figure 8-8 (later on in the chapter) shows this.

Variable expenses rise or fall with sales volume. These typically include truck expenses, travel and entertainment, advertising, and to a degree, repairs and maintenance. Fixed expenses are overhead costs that don't change with sales volume, like rent and utilities.

The Cash Flow Statement

This is the least understood of the financial statements, and is often omitted. But it does provide essential information. In the construction business, cash flow is critical to success. The owner should be intensely aware of changes in working capital.

A good level of working capital makes the difference between a healthy, thriving operation and one that's constantly struggling to stay one step ahead of the bill collectors. Working capital is the difference between current assets and current liabilities. As that level changes, so does the owner's ability to make a profit and grow.

The cash flow statement rearranges the information in the general ledger to show where cash came from and how it was spent during the year. The most obvious source for cash is profit. But this number has to be adjusted for non-cash expenses like depreciation. Cash also comes from loan proceeds and the sale of capital assets.

Cash is applied in several ways, too. You might buy new capital assets or pay off long-term debts. The net difference between the sources and applications of funds always equals the change in working capital.

On the cash flow statement, the sources and applications are listed first. Changes in working capital follow. The two totals will be equal.

There's no mystery to balancing the cash flow statement. Like the other financial statements, it's an arrangement of specific information in the general ledger which is already in balance. Figure 8-5 is an outline of a cash flow statement.

To prepare this statement, think of the general ledger as having three parts. First is the net profit for the year. That's always the first line on the cash flow statement. Net profit also accounts for all the balances in the income statement part of the general ledger. Remember, those are the income, expense and cost accounts.

Cash Flow Statement

SOURCES OF FUNDS	
Net Profit	________
Plus: Non-cash Expenses	________
Decreases in Long-term Assets	________
Increases in Long-term Liabilities	________
Total Sources of Funds	________
APPLICATIONS OF FUNDS	
Increases in Long-term Assets	________
Decreases in Long-term Liabilities	________
Total Applications of Funds	________
Net Change (Sources less Applications)	________
CHANGES IN WORKING CAPITAL	
Current Asset Changes	________
Current Liability Changes	________
Net Change	________

balance

Figure 8-5

The second segment includes all the balance sheet accounts except current assets and current liabilities. They're entered on the statement like this:

1) Changes in long-term assets account for a source of funds if an asset is sold, or an application if a new asset is purchased.

2) The increase in the accumulated depreciation account is a non-cash expense, and an adjustment to net profits.

3) Other increases in assets are applications of funds, and other decreases are sources. Assets increase when you add new capital, and decrease when you receive payment on a long-term note receivable.

4) Changes in long-term liabilities are a source of funds when you receive the proceeds of a new loan. If you reduce a long-term liability by paying off a debt, that's an application.

5) Changes in net worth by addition of money by the owner is a source of funds. And any draws are applications.

At this point, you'll have listed all of the changes during the year for long-term assets, long-term liabilities, and net worth. In the net worth section, your entry for net profit (or loss)

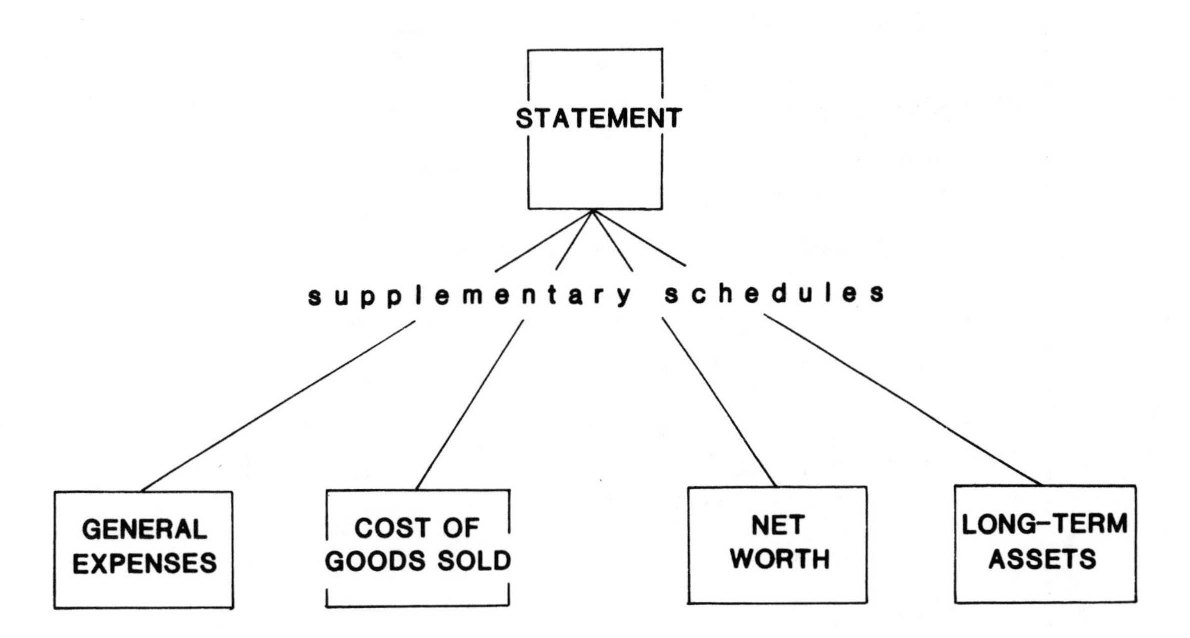

Figure 8-6

takes care of all the income and expense accounts. The balance at this point will be the increase or decrease in working capital.

Now, to prove that total, list all the changes during the year in current assets and current liabilities. Remember, when you add these accounts to the rest of the ledger accounts, the total will be zero. And since we're concerned here with change, the total of all changes will also equal zero.

The change you're reporting is simply a division of long-term and current accounts. Think of the balance sheet as having two segments. First is the change in each non-current account. These are accounted for in the sources and applications sections (including net profit adjusted for changes in depreciation, and increases and decreases in asset and liability accounts). The second part represents the amount of change — positive or negative — in all current asset and current liability accounts.

Here's how this works. You've already accounted for long-term assets, long-term liabilities and net worth. That total will always equal the changes in current accounts.

The last section of the cash flow statement, changes in working capital, summarizes this. If you now have more cash than you had at the end of last year, that's a plus to working capital. (Remember, cash is a debit-balance account.) Liabilities are the opposite. A higher balance now is a minus, because the liabilities are credit-balance accounts, and the opposite of assets.

The net sum of changes in current accounts will equal the net change in sources and applications of funds. Look ahead to Figure 8-11 for a completed cash flow statement.

Summarized Statements

At year end, or when applying for a loan, you'll need very detailed financial statements. At month end a more summarized version will probably do the job.

Sometimes it will just be more convenient to summarize significant totals and include details on attached supplementary schedules. This way it's easy to analyze financial statements quickly and see the status at a glance.

If someone needs to know more, they can refer to the supplementary pages. Figure 8-6 shows how the supplementary schedules support a

supplementary schedules

SCHEDULE 2: NET WORTH	
Owner's equity, January 1	$37,418
Plus: capital paid in	10,000
net profit	16,844
Owner's equity, December 31	$64,262

SCHEDULE 3: COST OF GOODS SOLD	
Inventory, January 1	$ 16,500
Plus: materials purchased	137,610
direct labor	149,011
other costs	14,392
Total	$317,513
Less: inventory, December 31	19,100
Cost of Goods Sold	$298,413

Figure 8-7

statement. One line on the statement is explained in detail in the supplementary report.

A summarized income statement would look like this:

Sales	$493,651
Cost of Goods Sold (SCHEDULE 3)	298,413
Gross Profit	$195,238
Variable Expenses (SCHEDULE 4)	– 32,416
Fixed Expenses (SCHEDULE 5)	–143,005
Operating Profit	$19,817
Federal Income Tax	2,973
Net Profit	$ 16,844

Compare that to Figure 8-10. You'll see that the bottom line is the same. All the important totals are there. The attachments contain the details of direct costs and fixed and variable expenses.

You can use the same approach for the balance sheet and the cash flow statement. The supplementary schedules would look like the ones in Figure 8-7.

You can summarize your financial statement in other ways, too. You can make a comparative statement that shows the results of one year's performance next to that for another year, and include the difference between the two years.

Maybe a comparison by percentages would be more useful. Or you could include both numbers and percentages on the same statement.

Comparison by percentage is most valuable if you want to observe trends or monitor particular accounts. On the income statement, dollar amounts for each account are usually followed by the account's percentage of total sales. You can see an example of this in Figure 8-8.

percentage breakdown statement

DESCRIPTION	AMOUNT	%
Sales	$493,651	100.0%
Cost of Goods Sold	298,413	60.5
Gross Profit	$195,238	39.5%
Variable Expenses	$ 32,416	6.5%
Fixed Expenses	143,005	29.0
Total Expenses	$175,421	35.5%
Net Operating Profit	$ 19,817	4.0%
Federal Income Tax	2,973	0.6
Net Profit	$ 16,844	3.4%

Figure 8-8

The Complete Financial Report

Include all three statements in every financial report. You can photocopy master forms and then prepare informal monthly reports by transferring information from the general ledger or trial balance.

Be sure to show whether or not the statement has been audited. If you send statements to your banker or other outsiders, you should let them know if your accountant hasn't reviewed the statements for accuracy.

Figures 8-9 through 8-11 make up a complete financial report that includes a balance sheet, income statement and cash flow statement.

ABC Construction Company
Balance Sheet
December 31, 1988

Current Assets:		
Cash in Bank	$ 13,902	
Accounts Receivable	79,632	
Inventory	19,100	
Total Current Assets		$112,634
Long-term Assets:		
Trucks	$ 37,400	
Equipment	18,945	
Furniture	10,462	
Total	$ 66,807	
Less: Accumulated Depreciation	(27,400)	
Net Long-term Assets		$ 39,407
Total Assets		$152,041
Current Liabilities:		
Accounts Payable	$54,380	
Payroll Taxes Payable	2,193	
Notes Payable	4,175	
Total Current Liabilities		$ 60,748
Long-term Liabilities:		
Notes Payable		27,031
Total Liabilities		$ 87,779
Net Worth:		$ 64,262
Total Liabilities and Net Worth		$152,041

Figure 8-9

ABC Construction Company
Income Statement
For the year ending December 31, 1988

Sales		$493,651
Cost of Goods Sold:		
Inventory, 1-1-88	$ 16,500	
Materials Purchased	137,610	
Direct Labor	149,011	
Other Direct Costs	14,392	
Total	$317,513	
Less: Inventory, 12-31-88	19,100	
Cost of Goods Sold		$298,413
Gross Profit		$195,238
Variable Expenses:		
Travel and Entertainment	$ 9,415	
Truck Expenses	14,837	
Repairs and Maintenance	5,984	
Advertising and Promotion	2,180	
Total Variable Expenses	$ 32,416	
Fixed Expenses		
Salaries and Wages	$ 86,000	
Payroll Taxes	7,942	
Rent	7,200	
Utilities	1,845	
Telephone	3,597	
Insurance	3,736	
Office Supplies	1,393	
Taxes and Licenses	3,909	
Legal and Accounting	2,900	
Printing	1,806	
Postage and Delivery	1,118	
Dues and Subscriptions	1,566	
Depreciation	9,300	
Maintenance	3,880	
Interest	1,683	
Bad Debts	4,200	
Miscellaneous	930	
Total Fixed Expenses	$143,005	
Total Expenses		$175,421
Net Operating Profit		$ 19,817
Federal Income Tax		$ 2,973
Net Profit		$ 16,844

Figure 8-10

ABC Construction Company
Cash Flow Statement
For the year ending December 31, 1988

Sources of Funds:		
Net Profit	$ 16,844	
Plus: Non-cash Expenses	9,300	
Total	$ 26,144	
Sale of Long-term Assets	13,406	
Total Sources of Funds		$ 39,550
Application of Funds:		
Purchase of Long-term Assets	$ 16,000	
Decrease in Notes Payable	10,406	
Total Applications of Funds		$ 26,406
Net Change		$ 13,144

Changes in Working Capital:

	Last Year	This Year	Net Change
Cash	$ 9,422	$13,902	$ 4,480
Accounts Receivable	68,124	79,632	11,508
Inventory	16,500	19,100	2,600
Accounts Payable	-49,433	-54,380	- 4,947
Payroll Taxes Payable	- 1,696	- 2,193	- 497
Notes Payable	- 4,175	- 4,175	0
Net	$38,742	$51,886	$13,144

Figure 8-11

Self Test

1. ***Use the trial balance on the following page to prepare a balance sheet as of December 31, 1988.***

2. ***Also prepare an income statement from the same trial balance.***

3. ***Prepare a cash flow statement from the trial balance.***

Trial Balance
December 31, 1988 and 1987

Account	1988	1987
Cash	$ 4,283	$ 13,902
Accounts Receivable	84,416	79,632
Inventory	17,400	19,100
Trucks	37,400	37,400
Equipment	18,945	18,945
Furniture	10,462	10,462
Accumulated Depreciation	– 36,300	– 27,400
Accounts Payable	– 42,811	– 54,380
Payroll Taxes Payable	– 2,701	– 2,193
Notes Payable, Current	– 4,175	– 4,175
Notes Payable, Long-term	– 16,550	– 27,031
Net Worth	– 64,262	– 47,418
Net Profit	6,107	16,844
Sales	– 473,010	– 493,651
Materials	124,840	137,610
Direct Labor	138,060	149,011
Other Costs	18,908	14,392
Change in Inventory	1,700	– 2,600
Variable Expenses	28,410	32,416
Fixed Expenses	153,907	143,005
Federal Income Tax	1,078	2,973
Net Profit	– 6,107	– 16,844

Chapter 9

HOW TO SET UP PAYROLL RECORDS

Payroll records are complex. But they return a lot of information that you have to retain and report to others outside your company. You have to keep track of your employees' pay and deductions, individually and overall. You also need to send that information to various government agencies on a regular basis.

Employees expect their pay on time. Your payroll accounting procedure has to be efficient and up to date for that to happen. It also needs to be simple and extremely "clean." Most company records are seen by no one other than the bookkeeper, the accountant, and the owner. On the other hand, payroll records are subject to inspection and audit by many outside agencies.

You can probably handle payroll records by hand if you have fewer than 20 employees. With a larger staff than that, you'd probably want to consider using a payroll service.

Payroll checks must be broken down among gross pay, several different deductions, and net pay for each employee. The details of that breakdown appear on the payroll check itself, the employee's record, and your checkbook or payments journal.

A pegboard (write-once) system is a convenient way to do payroll manually. On a pegboard, the check is written on top of the employee record and the permanent payroll summary record. All information transfers to all the records at once

When you use a bank or payroll service, you collect payroll information each period and send it by mail or phone for processing. The company then writes your checks and delivers them to you. They also prepare your quarterly and annual tax and information forms.

The Special Payroll Account

Payroll is one situation where a separate checking account may be worthwhile. First, you can use a special check format designed for payroll. You won't have to prepare a separate pay voucher to show the deduction breakdown.

An even better reason for a separate account is cash management. You have to make periodic tax deposits to federal and state agencies. The frequency of those deposits depends on the size of your payroll. You may also have to make other payroll-related payments such as workers' compensation, group insurance and union dues.

Deposit your total payroll costs each pay period, including benefits and employer contributions, to your special payroll account. Then when it's time to forward money for taxes and insurance, the funds will be there. If you use your own bank for payroll service, they can automatically transfer funds from your general account to the payroll account.

Another reason for a special account is posting convenience. The account breakdown for payroll checks is greater than most checks you write. Your disbursements journal would need several more columns just for payroll. And with many employees, your checkbook will be cluttered also.

With a separate account you'll have two payment journals, not just one. That means you'll have to post from one additional source each month. But you'll find this extra step will save time and effort in the long run.

Types of Payroll Taxes

Whether employees are paid by the hour or a fixed salary, you have to withhold several deductions.

1) FICA is the social security tax. The initials stand for Federal Insurance Contributions Act, which is the law establishing payroll withholdings for social security. The amount withheld is a percentage of each employee's pay, up to a maximum set each year.

2) Federal Income Tax (FIT) is based on the level of pay, whether the employee is married or single, and his or her claimed exemptions.

3) State Income Tax (SIT) if applicable, is similar to FIT.

4) State Disability Insurance (SDI) is collected by most states as a percentage of gross pay.

Besides the withheld deductions, employers must pay these additional taxes:

1) FICA - the employer matches the amount withheld for each employee.

2) FUTA - Federal Unemployment Tax Act. Employers pay into a federal unemployment insurance fund based on employees' gross pay.

3) SUI - State Unemployment Insurance, if applicable to your state, is similar to FUTA.

4) Besides payroll taxes, various union assessments may apply to certain employees. In some cases, the benefit is taxable. If so, it must be added into gross pay before taxes are computed, and then deducted again. This increases payroll taxes. In other cases, union benefits are provided by the employer and paid under a separate procedure. In that case, payroll records are not involved.

Payroll Records

Your payroll records must let you do four things:

1) You have to keep track of each employee's pay and deductions.

2) You need to know the employer's liability and include payments with periodic payroll tax deposits.

3) You need accurate year-to-date records for each employee, because some withholdings stop at a certain point.

4) You need information to prepare payroll tax deposits and quarterly and annual tax returns.

First, you need an employee payroll record that contains information like the form in Figure 9-1.

Record the employee's name, address, and social security (taxpayer's identification) number. You *must* have the social security number for your reports, so be sure to get it from your employee. You're subject to fines if you don't furnish the social security number.

"Status" refers to the employee's marital status. The three types are: married, single, and

Employee Payroll Record

name ____________________ soc.sec. # ____________

status ________ exemptions ____________ year ____________

$ ________ per ________

DATE	GROSS	FIT	FICA	SIT	SDI	TOTAL	NET
		DEDUCTIONS					

Figure 9-1

head of household. A married employee may elect to be taxed as a single person when he wants more taxes deducted from his paycheck. "Exemptions" is the number of dependents the employee is claiming. The status and exemptions lines tell you which withholding chart to use in computing federal and state income taxes.

Payroll is always reported quarterly and for the calendar year January 1 to December 31 regardless of your business tax year, so include the year on the employee payroll record.

You also need the employee's pay rate and period. For example, you might pay one employee $15.80 per hour, while another gets $3,000 per month.

List each paycheck on the permanent record. There should be columns for the following:

- The check date
- Gross pay before deductions
- Each deduction
- Total deductions
- The net amount of the check

Your line-by-line breakdown might also include space for the check number, the number of hours worked, and the payroll period covered.

If you pay union benefits, tax them, and then deduct them, you'll need still more columns. Most stock payroll forms allow more columns than shown in Figure 9-1.

the payroll checking account

NAME	CHECK	DATE	GROSS	DEDUCTIONS					NET
				FIT	FICA	SIT	SDI	TOTAL	

Figure 9-2

This record is essential and must be complete. It's required by the government. You must record the amounts you pay and withhold for your employees, for the I.R.S., your state tax agency, and your workers' compensation insurer.

Next, you need a payroll check register. You list and account for each check in order by date and employee name. Then you list the same items shown on the individual employee's record. Figure 9-2 shows a sample check register/payroll journal.

Be sure to account for every check, and always in numerical order. If you void a check, write in the number and the date, write "void" in the name column, and a zero in the net column. This journal is the book of original entry for payroll records. You'll balance it each month just like your general account.

The Payroll Worksheet

At the end of each quarter, you have to file a payroll tax return for the federal (and if applicable, state) government. To do this, you need to know the year-to-date gross pay for each employee. You can use a payroll worksheet like the one in Figure 9-3 for this.

Payroll Worksheet

date ____________

name	previous gross	gross	FICA		SDI		SUI	
			under 45,000	over 45,000	under 21,900	over 21,900	under 7,000	over 7,000

Figure 9-3

Here's an example. Let's say that in your state, disability insurance is collected on wages up to $21,900. That means that once an individual's pay exceeds that ceiling, you no longer deduct that tax. The same limitations apply to state unemployment insurance and to FICA.

You need to know the previous year-to-date gross and the current gross pay to identify those ceilings. Each employee's situation is different with regard to tax withholding. Here's how the first three columns of the worksheet will look filled out:

Name	Previous Gross YTD	Current Gross
Bob Adams	$ -0-	$1,500
Mark Carter	6,300	1,200
Dan Mackey	21,450	1,200
Bill Smith	22,500	2,100
Ted West	43,900	3,000

In the example, Bob Adams is a new employee who is under all of the ceilings. All of the deductions apply.

Mark Carter is under the FICA and SDI limits. But part of his current pay puts him over the SUI limit of $7,000. Only $700 of the current check is subject to this tax ($7,000 limit minus $6,300 previous gross).

Dan Mackey is over the maximum on SUI, so that tax no longer applies for this year. For SDI, only a portion of the current paycheck is subject. The maximum is $21,900, and his previous pay is $21,450. So only $450 of the current check is subject to SDI. The rest is over. Mackey is still below the FICA ceiling.

Bill Smith is over both SDI and SUI ceilings, but under the FICA. Only the FICA tax is withheld.

Ted West is over both SDI and SUI, and neither tax applies for the rest of the year. And only part of the current paycheck is subject to FICA withholding. That part is the difference between the ceiling and the previous gross, or $1,100 ($45,000 less $43,900). The balance, $1,900, is over the maximum and not taxed. In the next pay period, he'll be over the maximum in all three categories.

Now you can fill out the worksheet. In each set of two columns, you divide the current gross between the "under" and "over" headings. Whether the tax is deducted from the employee's check or paid by the employer, you can easily see how much pay is subject to each type of tax for each employee. You will avoid over-deducting or over-paying the taxes.

Computing Payroll Tax Liabilities

At the end of each month, you post a journal entry for your employer's portion of taxes and to record payroll in the general ledger.

Remember, there are two types of payroll taxes: those that are deducted from the employee's check, and those the employer pays.

The first type, deductions, are withheld and then sent to the government. There is no expense entry. You set up a liability when you write the payroll checks and then reverse it when you forward the money to the government.

The second type, paid by the employer, are business expenses. You set up a journal entry to recognize the payroll taxes expense, and also to increase the liability. The employer's portion is deposited to the appropriate agency as often as your payroll size requires. Figure 9-4 shows how these two separate liabilities are recorded.

You record the liability for taxes withheld from the payroll account journal. The entries below shows the breakdown of the entire payroll for the period.

Explanation	Debit	Credit
Direct Labor	27,954.60	
Salaries and Wages	4,500.00	
FICA payable		2,320.50
FIT payable		4,594.10
SDI payable		519.27
SIT payable		1,298.18
Cash		23,722.55

Explanation	Debit	Credit
Payroll Tax Expense	3,683.60	
FICA payable		2,320.50
SUI payable		1,103.46
FUTA payable		259.64

payroll taxes

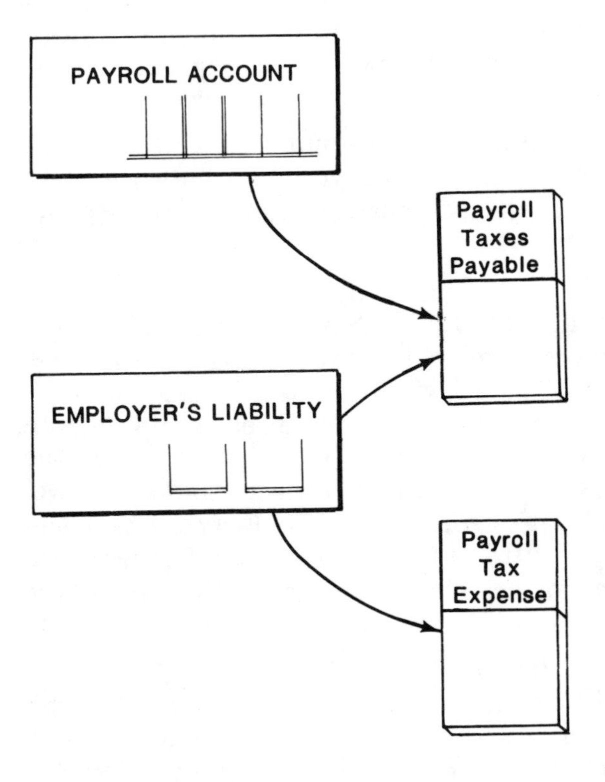

Figure 9-4

In most cases you won't prepare a separate journal like this one. You'll post directly from the permanent payroll account disbursements total or from a summary you'll get from your payroll service. However, be sure you understand how the entry is broken down.

The general payments journal consists of a number of debits to various accounts, and one credit to the cash account. The payroll payments journal is different. The debits are to direct labor (a direct cost) and salaries and wages (an operating expense). They are offset by credits to the various payroll liability accounts and to cash.

This entry takes care of the liability for the amounts withheld. You still need to record the employer's share. That requires a journal entry, like this:

The employer's share of FICA is the same as the amount withheld for the period. SUI is strictly an employer-paid tax, and is assessed on all payroll up to the individual maximum. And FUTA is also a tax just on the employer, based on gross payroll.

The Liability Account

With entries from the payroll register and from a journal entry, the activity in the liability account is heavy. Also, you need a breakdown for each type of tax withheld or paid. You'll need a different format for the payroll liability account in the general ledger.

The liability increases each time you do a payroll. It decreases when deposits are sent to the government. An increase in a liability account is a credit, and a decrease is a debit.

Figure 9-5 shows a Payroll Taxes Payable liability account. The credits (increases) to this account come from various journal entries, and are further broken down in sub-accounts on the left. When you make deposits and quarterly returns, you'd reduce these levels with debit entries.

Here's a summary of guidelines for posting the payroll liability account.

1) The liability account increases for each payroll.

2) You have to compute employer's liability for each pay period. The liability is computed based on current rates and maximums for each type of tax. Use the payroll worksheet to figure the taxes.

3) The liability account is reduced (debited) each time you forward money to the government agencies.

The total of the liability sub-accounts must always equal the amount you currently owe for payroll taxes.

Government Forms

There are a number of forms and filing requirements at the federal level, and others for reporting state taxes. You can ask your accountant what you have to do, when you have to do it, and how to fill out the forms. There are also instruction books available from the I.R.S, the Social Security Administration and state agencies.

Here's a summary of the most common forms:

W-4, Employee's Withholding Allowance Certificate — must be filled out by every employee, and filed permanently by your company. The employee figures out how many exemptions to claim, lists his social security number, correct name, and full address. Be sure to use the current revision of the W-4. Employees may change their W-4 at any time, and may be required to refile each year.

On the back of the W-4 there are two worksheets. One is for deductions and adjustments, the other for two-earner/two-job calculations.

W-2, Wage and Tax Statement — filed at the end of each year with the Social Security Administration. You use this form to summarize each employee's pay for the year.

The W-2 is a six-part form. The original goes to the Social Security Administration. Copy 1 is for your state or local taxing agency. Copies B, C, and 2 are for the employee to use in filing income tax returns. The last copy is for your files.

W-3, Transmittal of Income and Tax Statements — sent to the Social Security Administration with their copy of the W-2 forms.

You use the W-3 to summarize and reconcile the W-2 forms before you distribute them. It includes the address for mailing the original W-2's.

Before you begin paying payroll taxes, you must apply to the I.R.S. for an employer's identification number. Once you've received that, the forms you need will be mailed to you. However, if you need extras, or you're setting up payroll records for the first time, you might need to order forms on your own. To do this, use Form 7018, Employer's Order Blank for Forms. The back side of 7018 lists the addresses for ordering forms in each state. You can get this form from the Internal Revenue Service.

Form 8109-B, Federal Tax Deposit Coupon — used to make payroll tax deposits during the quarter. You complete the coupon and deliver it to any Federal Reserve Bank with your check.

Payroll Taxes Payable

SUB-ACCOUNTS											
FICA	FIT	FUTA	SDI	SIT	SUI		DATE	REF.	DEBIT	CREDIT	BALANCE
—	—	<63.50>	<406.19>	<416.00>	<157.43>	Sub-total					<1,043.12>
<2,320.50>							3-15	J3-1		2,320.50	
	<4,594.10>						3-15	J3-1		4,594.10	
			<519.27>				3-15	J3-1		519.27	
				<1,298.18>			3-15	J3-1		1,298.18	
<2,320.50>							3-15	J3-2		2,320.50	
					<1,103.46>		3-15	J3-2		1,103.46	
		<259.64>					3-15	J3-2		259.64	<13,458.77>
<4,641.00>	<4,594.10>	<323.14>	<925.46>	<1,714.18>	<1,260.89>	Sub-total					

Figure 9-5

Form 8109-A, Instructions — detailed description tells you how to fill out the deposit coupon. It includes an order form for deposit coupons. Once you've established your account, the deposit forms will be sent to you with your federal employer i.d. number, name and address printed on them.

Form 941, Employer's Quarterly Federal Tax Return — used to report payroll for the period, reconcile deposits and pay the balance of the taxes due.

You list the total of all wages paid, then report the total amounts withheld for income tax. The calculation of the FICA liability includes the amount withheld *and* the matching employer's contribution.

If your quarterly liability is more than $500, you have to break down details of your payroll tax liability for each payroll period during the quarter. When you do this, be sure you list the *liability* for each period, not your deposits.

If you made a mistake on a previously filed Form 941, you use 941-C to correct it. You can get this form from your local Internal Revenue Service office.

Note that most business records can be discarded after a certain time following filing of tax returns. This isn't the case with payroll records, which you should keep indefinitely.

Self Test

1. *The detailed information about gross pay, deductions, and net pay, must be shown on:*

a) the stub of each payroll check.

b) the employee's record.

c) the payroll checking account.

d) all of the above.

2. *The payroll liability account is used to:*

a) account for all payroll tax deposits.

b) record the amount due for withholding and for the employer's taxes due.

c) keep track of what is due to each employee.

d) all of the above.

3. *The employee's payroll record serves the purpose of verifying:*

a) the amounts paid and deducted, and documenting payroll in the event of an audit.

b) the hours each employee works, so that payroll can be correctly calculated.

c) that the payroll account is in balance before checks are issued.

d) all of the above.

4. *Having a special payroll account is a convenience because:*

a) the breakdown of each check has to be reported to every employee.

b) the distribution of the net check is more detailed than for most types of payments.

c) it lends itself well to write-once systems, so that you avoid duplication.

d) all of the above.

5. *The liability for payroll taxes is computed as:*

a) the amount withheld from each employee's check, only.

b) the rate charged for each type of tax.

c) the combination of withholdings and the employer's portion of tax.

d) the total of all deposits made for the quarter.

6. *The maximum levels of various payroll taxes are:*

SUI — $ 7,000

SDI — 21,900

FICA — 45,000

This month's payroll consists of:

Employee	Previous Gross	Current Gross
A	$ 5,500	$1,000
B	6,500	1,200
C	19,500	2,500
D	24,600	2,500
E	42,000	4,000

Break down each employee's current gross into over/under categories, using the worksheet below.

Employee	SUI - $ 7,000		SDI - $21,900		FICA - $45,000	
	under	over	under	over	under	over
A						
B						
C						
D						
E						
Total						

CHAPTER 10

RECORDS FOR FIXED ASSETS

The tax rules say you can deduct legitimate business expenses from income. That way, you reduce the income you'll have to pay taxes on. But when you buy a capital asset, you have to spread the deduction over several years. That's called depreciation.

A capital asset is tangible property that has a useful life of more than one year. If you spend $85 this year for operating supplies, that's an expense. But if you also spend $8,500 for a piece of equipment, that can't be written off as a business expense. It must be depreciated.

In this chapter you'll see how to figure depreciation. I'll also describe the records you need to keep for capital (also called *long-term* or *fixed*) assets.

If your job is limited to bookkeeping, you don't need to decide what is or isn't a capital asset. Nor will you have to determine which depreciation method to use, or how that decision affects taxes. Your accountant will usually make those decisions. My purpose here is to give you a solid understanding of how fixed assets are handled in the books, and how to keep track of the depreciation your accountant decides to claim.

Depreciation Methods

Depreciation is nothing more than a deduction for part of an asset's cost. Unlike the operating supplies you buy and consume every year, an asset will be used for many years. Consequently, the asset's cost is written off over many years.

Before tax reform, you were allowed to estimate the number of years in an asset's useful life. Now the tax rules have defined a limited number of recovery periods. Assets are depreciated based on the recovery period to which they belong. And the tax laws restrict which of the many depreciation methods you're allowed to use.

Depreciation is calculated in one of two ways. First is the *straight-line method*. That means that the same amount is depreciated each year. If an asset belongs in the 10-year recovery period, and you use the straight-line method, you'll deduct one-tenth of the asset's cost each year.

The formula for straight-line depreciation is the easiest of all methods. You divide the basis

depreciation methods

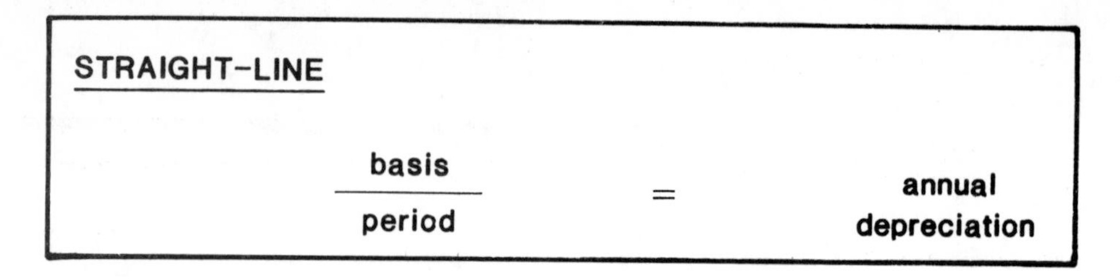

150% DECLINING BALANCE

$$\frac{\text{basis} - \text{prior depreciation}}{\text{period}} \times 150\% = \text{annual depreciation}$$

Figure 10-1

(usually the cost of the asset) by the number of years in the period. The answer is the amount of depreciation you'll claim each year.

The second method is the *declining balance.* Under this method, you calculate the straight-line rate. Then you increase the depreciation amount by a specified percentage. The actual deduction you claim goes down each year.

Suppose you buy an asset worth $8,000 that you can depreciate over five years. That means that you would claim a deduction of $1,600 each year if you used the straight-line method.

If you decide to use the declining balance method, you'll depreciate that same asset faster during the early years. If you use the rate of 150 percent, the first year's depreciation will be 150 percent of the straight-line rate.

You start with the basis (cost) and then figure the straight-line rate:

$$\frac{\$8{,}000}{5} = \$1{,}600$$

Next, you multiply the straight-line depreciation by the declining balance percentage:

$$\$1{,}600 \times 150\% = \$2{,}400$$

After the first year's depreciation has been claimed, your basis in that asset has been reduced. The book value (cost minus depreciation) is now $5,600:

$$\text{cost} - \text{depreciation} = \text{net basis}$$

$$\$8{,}000 - \$2{,}400 = \$5{,}600$$

In the second year, you would compute depreciation the same way, but this time using $5,600 as the basis:

1) Compute straight-line rate:

$$\frac{\$5{,}600}{5} = \$1{,}120$$

2) Multiply straight-line depreciation by the declining balance rate:

$$\$1{,}120 \times 150\% = \$1{,}680$$

After the second year, the net basis is reduced to $3,920:

$$\$5{,}600 - \$1{,}680 = \$3{,}920$$

This will be used to compute the following year's depreciation. Figure 10-1 summarizes the two depreciation methods.

Claiming Depreciation Today

Today's depreciation tax rules are known as the *Accelerated Cost Recovery System* (ACRS). They grew out of the Economic Recovery Tax Act of 1981, and have been modified since then by a number of other tax laws.

Depreciation for most assets under the ACRS system is a combination of straight-line and declining balance depreciation. If your accountant advises using the prescribed ACRS method, you'll begin depreciating assets on the declining balance method, and revert to straight-line in the final years.

In Figure 10-2, you can see the schedule for reverting to straight-line depreciation. In the ten year depreciation class, 200 percent declining balance is applied to years one through six. The remaining years are depreciated on the straight-line basis.

The ACRS recovery classes and allowed depreciation methods are:

3 years	200% declining balance
5 years	200% declining balance
7 years	200% declining balance
10 years	200% declining balance
15 years	150% declining balance
20 years	150% declining balance
27.5 years	straight-line only
31.5 years	straight-line only

The 27.5 year class is for residential rental property only, and the 31.5 year class is for commercial real estate. In these classes, you have no choice but the straight-line method. Also, land can't be depreciated under any circumstances. Only the value of buildings and other improvements can be depreciated.

In all other classes, you can claim depreciation under the prescribed method, or elect to use straight-line rates. The prescribed methods begin with declining balance, and then revert to straight-line. In the first year, the allowed deduction is one-half of the full year's allowed depreciation. A specific percentage of the asset is depreciated each year, as shown in Figure 10-2.

	Percentage Depreciated					
Year	**3 year**	**5 year**	**7 year**	**10 year**	**15 year**	**20 year**
1	33.00	20.00	14.28	10.00	5.00	3.75
2	45.00	32.00	24.49	18.00	9.50	7.22
3	15.00	19.20	17.49	14.40	8.55	6.68
4	7.00	11.52	12.49	11.52	7.69	6.18
5		11.52	8.93	9.22	6.93	5.71
6		5.76	8.93	7.37	6.23	5.28
7			8.93	6.55	5.90	4.89
8			4.46	6.55	5.90	4.52
9				6.55	5.90	4.46
10				6.55	5.90	4.46
11				3.29	5.90	4.46
12					5.90	4.46
13					5.90	4.46
14					5.90	4.46
15					5.90	4.46
16					3.00	4.46
17						4.46
18						4.46
19						4.46
20						4.46
21						2.25

Figure 10-2

Each class takes one additional year, since only half of the computed depreciation can be claimed during the first year.

These rates are based on the Tax Reform Act of 1986, and might change with future federal tax legislation.

Elective Methods

Besides using the prescribed depreciation methods, you can also elect to use straight-line for assets belonging to one class put in service during the year. This election must be applied to *all* assets in that class, and the election is made each year.

Straight-line rates are:

- 5 years: cars and light, general-purpose trucks.
- 9.5 years: computer-based telephone switching equipment.
- 12 years: personal property with no class life category.
- 40 years: real estate.

The last category is an alternative period for claiming depreciation on real estate, which otherwise is depreciated over 27.5 years (residential) or 31.5 years (commercial).

Another election is to expense capital assets. The law now says you can immediately write off or depreciate up to $10,000 in new assets.

There is a limit to this rule. If you buy more than $200,000 in assets during a single year, the expensing provision is reduced one dollar for each dollar over $200,000.

Here's how this works. Suppose this year your company buys $206,550 in new assets. The expensing provision is reduced dollar for dollar by the excess over $200,000:

$10,000 – $6,550 = $3,450

The most you can expense this year is $3,450.

The decision to expense, claim straight-line depreciation, or use the prescribed method, depends on your estimated current and future tax liabilities. If you need higher deductions this year, you'll want to claim the maximum depreciation allowed. You should discuss these decisions with your accountant.

Posting Depreciation

Depreciation is strictly a non-cash journal entry. First you buy a capital asset with your own or borrowed money. Then each year, you record the depreciation you're allowed to claim.

Usually, you'll only record depreciation at year-end as a closing adjustment. But you should estimate it every time you prepare a financial statement if the amount to be depreciated is substantial.

The journal entry consists of a debit to depreciation expense, and a credit to the asset account, "accumulated depreciation" (also called "reserve for depreciation" in some companies):

	Debit	Credit
Depreciation Expense	XXX	
Accumulated Depreciation		XXX

The debit increases expenses, so it also reduces profit. And the credit lowers the book value of assets. This is illustrated in Figure 10-3.

When you prepare year-end financial statements, you normally show total depreciation as a single item. You'd list each type of asset, then deduct the total of accumulated depreciation to arrive at net assets. Your balance sheet would look like this:

Furniture and Fixtures	$ 18,403
Trucks and Autos	62,900
Machinery	38,352
Building	62,500
Land	40,000
Total	$222,155
Less: Accumulated Depreciation	45,018
Net	$177,137

You can also use a depreciation schedule to show the accumulated depreciation in each category. Remember that the asset account is the total of all depreciation claimed — not only this

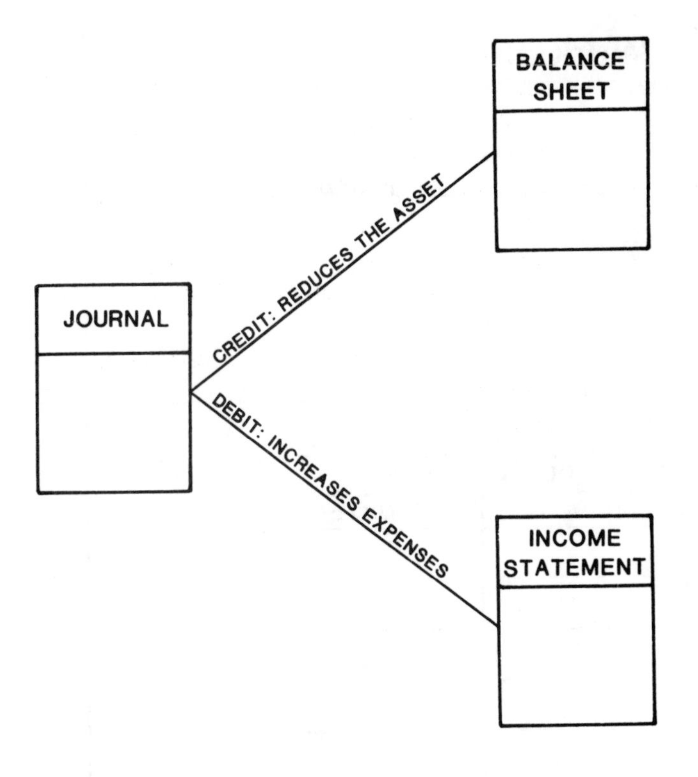

Figure 10-3

year, but for as long as assets have been on the books.

The schedule format is shown below.

	Gross	Accumulated Depreciation	Net
Furniture and Fixtures	$ 18,403	$ 4,660	$ 13,743
Trucks and Autos	62,900	22,900	40,000
Machinery	38,352	10,043	28,309
Building	62,500	7,415	55,085
Land	40,000	—	40,000
Total	$222,155	$45,018	$177,137

Use a worksheet like the one in Figure 10-4 to keep track of depreciation for each asset. I recommend that you assign an identification number to each asset. You can use that number to keep track of expenses and maintenance costs related to a particular asset.

"Class" refers to the type of asset: furniture, equipment, etc.

Use a worksheet like this one to keep the accumulated depreciation account in balance.

Asset Records

Besides tracking depreciation, you should also keep an eye on the asset itself. This is especially true if you have many capital assets or if you buy and sell them often. Use a form like the one in Figure 10-5 to collect asset information.

This information is also in the general ledger. But you can refer to it more easily if it's all in one place. You won't have to look up individual invoices. Fill out one of these whenever you buy or sell a capital asset.

You also want to keep usage records for assets. As an asset ages it will need more maintenance. Idle time and repair costs will increase until you'll want to consider replacement. Figure 10-6 is a sample use record for assets.

You'd use a form like this for equipment used on the job, like machinery and trucks, but not for furniture or buildings. Fill in the scheduled and actual hours the equipment is operated, and idle. Account for all the time. You'll be able to see right away when idle time increases beyond an acceptable level due to repairs.

You should also keep asset cost records similar to the one in Figure 10-7. Use it to spot trends in maintenance and operating costs. You'll be able to tell when it becomes cost-effective to replace a certain asset.

Some types of equipment are maintained constantly, and not actually replaced, or are replaced part by part. For example, you might restore the engine on an expensive piece of equipment regularly over a period of time.

If replacement cost is high, usage and maintenance records are essential. It may appear that on-going maintenance is less expensive than replacement. These records might prove otherwise.

Depreciation Worksheet

Asset ______________________________ Number __________

Class ______________________________

Depreciation method ______________

Cost ______________

Purchase date ______________ Sale date ______________

YEAR	BASIS	DEPRECIATION	
		CURRENT	TO DATE

Figure 10-4

Asset Record

Description ______________________________

Class ____________________ Number ________

__ New __ Used Purchase date ____________

Price	$	______
Tax		______
Delivery		______
Installation		______
Maintenance		______
Other		______
Total	$	______

Location of asset ______________________

Sale date ____________

Price	$	______

Figure 10-5

Use Record

Month/Year ________

Description ______________________________ Number ________

Day	HOURS:		Job #	HOURS:	
	From:	To:		operated	idle

Figure 10-6

Asset Cost Worksheet

Description ______________________________________ **Number** __________

Date	Labor	Parts	Fuel	Maintenance	Total

Figure 10-7

Self Test

1. ***Depreciation is:***

    a) a reserve for future repair and maintenance costs.

    b) the actual wear and tear on capital assets.

    c) a yearly deduction for the cost of an asset.

    d) all of the above.

2. ***Straight-line depreciation is:***

    a) the required depreciation method for real estate.

    b) allowed as an alternative method of depreciation.

    c) the easiest method to compute.

    d) all of the above.

3. ***Declining balance depreciation:***

    a) provides a higher deduction in the earlier years.

    b) provides a lower deduction in the earlier years.

    c) is no longer allowed under tax rules.

    d) is allowed only on real estate.

4. ***The depreciation journal entry consists of:***

    a) a debit to the asset account, and a credit to accumulated depreciation.

    b) a debit to the expense account, and a credit to profit and loss.

    c) a debit to the expense account, and a credit to accumulated depreciation.

    d) no actual entry, since it's a non-cash calculation.

5. ***Your company owns its own building. Depreciation must be calculated using:***

    a) the 10-year class.

    b) declining balance.

    c) 27.5-year straight-line.

    d) 31.5-year straight-line.

6. Your company purchases an asset for $5,000. Calculate depreciation, assuming it belongs in the 5-year class, under both the prescribed method and the elective straight-line method. Use the worksheet below:

Year	Prescribed Method	Straight-line Depreciation
1		
2		
3		
4		
5		
6		

CHAPTER 11

RECORDS FOR ACCOUNTS RECEIVABLE

Accounts receivable often has a high volume of activity. You need to post every time you bill a customer or receive a payment. Subsidiary records are useful here. They'll remove detail and clutter from your general ledger and will make posting and balancing easier.

Your accounts receivable system will:

1) Contain the details that are summarized into the general ledger each month.

2) Hold up-to-date information about each customer account so you can send out accurate monthly statements and collect delinquent accounts.

The Subsidiary System: An Overview

The posting chain doesn't change with a subsidiary system. Source documents support and prove what you post onto journals, and journals are posted into the ledger.

Accounts receivable is a subsidiary of the journal and ledger functions. It's a tedious process to post a large volume of charges and payments each month to an income journal. And the information won't be organized conveniently.

If you bill for most of your jobs, you'll only need an income journal for miscellaneous cash sales.

You'll post the details of most of your income in a subsidiary accounts receivable account.

In this account, you'll set up one account record for each customer. On that record you'll post all charges and payments for that one customer. You'll be able to review the status of each account and send out monthly statements with the information contained in the account record cards.

The account records are filed alphabetically by customer to make posting easier.

If you only have 20 or fewer transactions a month, you can post all charges and payments to a master record called the accounts receivable ledger. This is a variation of the income journal and holds only the accounts receivable transactions.

To be sure your posting is accurate, you'll first develop a control by adding up your billings. Then you'll do the same with receipts. You'll use those totals to check the amounts posted to the accounts receivable ledger and the individual account cards.

Then you'll post the transactions. A comparison between the control total and what you've posted will prove the accuracy of your work.

At the end of the month, you'll enter a summary of the accounts receivable activity into the general ledger before you close the books. You'll also send statements to your customers who have open balances.

Figure 11-1 shows how the parts of the subsidiary system relate to each other.

Once your posting volume increases, you'll have to find a more efficient posting method. You can save time by using a write-once or automated system. But you'll still need to check what you've posted against a control.

Posting Accounts Receivable

Develop a routine method for posting accounts receivable so your system will run efficiently.

Always use the same sources to post your accounts receivable system. This will help you avoid mistakes. Prepare invoices for all of your charges and use them as source documents for the accounts receivable subsidiary ledger. The information you'll need from each invoice is the date, customer's name, invoice number and amount of the sale.

Just as the invoices serve as source documents for charges, bank deposit information should serve as the source document and control figure for payments.

In one month you might send out 45 bills that you prepare in weekly batches. Before you begin to post, add the total of all the invoices for the week. That's your control total. Now post the invoices to the ledger and to the account cards. A pegboard system would be a time-saver in this case.

Next, you'd total the posted entries and compare that total to your control number. If the numbers don't agree, find and correct the errors. Balance the subsidiary system before you do any more posting.

The details posted on customer account cards always represent the sum of everything posted on the accounts receivable ledger. You'll know if there's an error in either the accounts receivable ledger or an account card by comparing the totals to the control. But that's only one form of control. You must also be sure that the amounts posted to the accounts receivable ledger and the cards agree with the invoice amounts.

If you post an invoice to the ledger as $2,895 and to the card as $2,859, you'll know there's a mistake because the daily posting in the ledger won't agree with the daily posting total for the account cards.

You could post the wrong amount to both the ledger and the account card. This would show up when you compare totals to the control figures for the day. You get the control figures by adding your source document totals, in this case the invoices for the period.

Here's a review of the accounts receivable posting sequence for a manual system:

1) Run an adding machine tape of your source documents. These are invoices or checks received. This batch total is your posting control.

2) Post the accounts receivable ledger. The posting control total must agree with the total posted to the subsidiary ledger. Check this balance before going on.

3) Post the entries from the source document (invoice or receipts journal) to the customer account cards. You might do

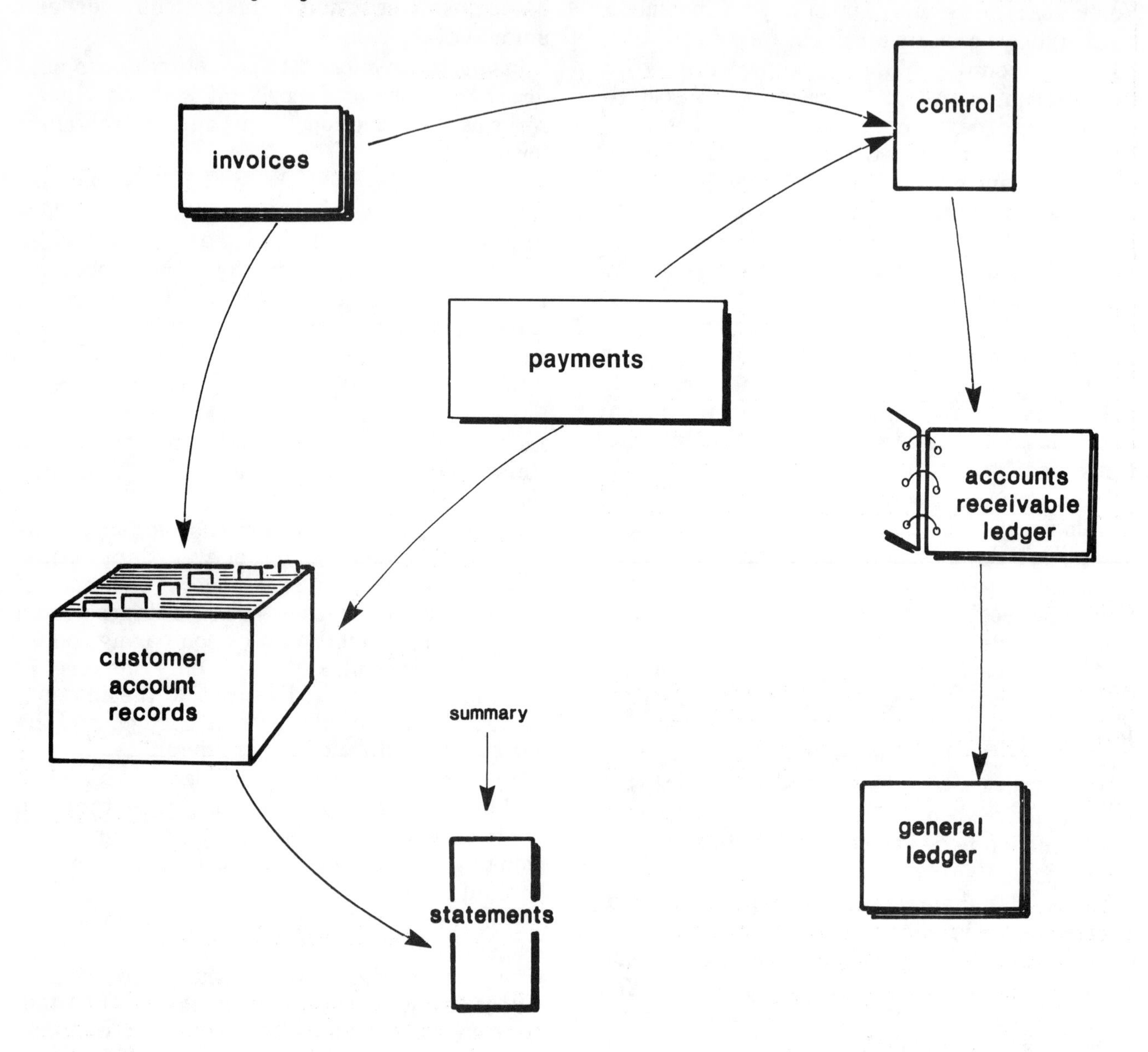

Figure 11-1

this daily, weekly or monthly, depending on the volume.

If you have very little accounts receivable activity, you can enter a transaction to the ledger and account cards at the same time. This way, the customer accounts are always current. Again, be sure the total of the entries posted to the account cards agrees with the control.

4) At the end of the month you must balance the entire subsidiary system. You do that by running an adding machine tape of all the customer account balances. That total should be same as the balance on the accounts receivable control and ledger.

Do this before you send out statements. If there are errors in accounts, your statements will be wrong. That's bad for your credibility. And it can lead to collection problems.

Suppose you send a bill to a customer with a math error in his favor. Then you send a corrected bill. The customer probably won't say

Control

MONTH	CHARGES	PAYMENTS	BALANCE

Figure 11-2

anything when he gets the original, favorable bill. Later, though, you'll get an angry reaction if the amount due is much higher.

Here are some of the mistakes that can put your books out of balance:

1) Posting a charge as a payment, or a payment as a charge.

2) Making an addition or subtraction error on the customer account or on the accounts receivable ledger.

3) Posting the wrong amount.

4) Omitting an entry.

5) Beginning a new customer account card with an incorrect balance forward.

6) Posting to the wrong account. In this case you would still be in balance. The only way to avoid this problem is to be especially careful when posting. Sort your source documents alphabetically and by date before you post.

This system will work as long as your accounts receivable volume is fairly low and consistent. But when your volume increases, you'll save time and eliminate duplication by converting to a write-once or fully automated accounting program.

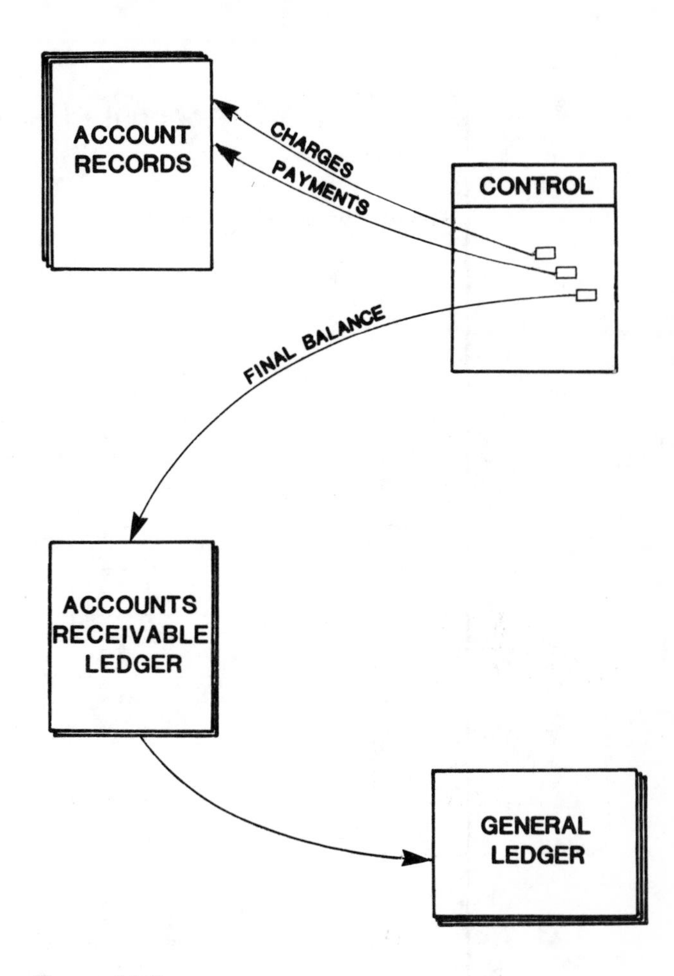

Figure 11-3

The Accounts Receivable Control

The control card is the crucial record to ensure accuracy in your subsidiary system. It summarizes all the correct totals for the month. It shows the previous balance forward, all charges and payments for the month, and a final balance. A sample control card is shown in Figure 11-2.

Its final balance must agree with the balance on the accounts receivable ledger, and with the general ledger accounts receivable account. That balance will also be the sum of the ending balances of all the customer account cards. The relationship between the control, the customer accounts, the accounts receivable ledger and the general ledger is illustrated in Figure 11-3.

If your receivables volume gets heavy, it may be necessary to expand the control feature by setting up logs for listing invoices and cash re-

Accounts Receivable Ledger

DATE	DESCRIPTION		CHARGES	PAYMENTS	BALANCE

Figure 11-4

ceipts. These will contain details of the summaries entered to the control card.

The Accounts Receivable Ledger

Figure 11-4 is the format for an accounts receivable ledger. The date column will contain the transaction date. The description will usually be an invoice or check number. The small column is for a posting check when information is transferred to the customer account. Amounts billed or received go in their respective columns.

You might choose to post just the batch control total to the subsidiary ledger if you have more than a few transactions each day. In that case the control would still be effective. Even if you make a mistake in running the tape of the source documents, the chance of making the same mistake in posting from the source documents to the account cards is very unlikely. You'd know there's a mistake by comparing the totals of the control to the account postings.

account record

customer ______________________________

address ______________________________

____

contact ______________ telephone __________

terms ______________ credit limit __________

DATE	INVOICE	CHARGES	PAYMENTS	BALANCE

Figure 11-5

The Account Record

The customer account card is a perpetual and cumulative record. The balance forward is carried from one month to the next, and each charge and payment is entered as it occurs.

Figure 11-5 is an example of a customer account card. Include payment terms (such as any discount you allow) and the customer's credit limit.

The body of the form is the transaction record. Show date, invoice or check number, amount of each charge or payment, and the balance.

Sample Entries

Let's go through a few sample entries to see how the various records relate to one another. At the beginning of May, one contractor's accounts receivable control looks like this:

Control			
Date	**Charges**	**Payments**	**Balance**
4-30			74,655.00
5-8	30,202.00	24,200.00	
5-8			80,657.00

His accounts receivable ledger for the first week looks this way:

Accounts Receivable Ledger				
Date	**Description**	**Charges**	**Payments**	**Balance**
4-30	Balance forward			74,655
5-4	Smith, inv. 878	1,095		75,750
5-4	Adams, paid		11,350	64,400
5-4	Jones, paid		350	64,050
5-5	Smith, paid		500	63,550
5-5	Marks, inv. 879	15,000		78,550
5-5	West, inv. 880	8,000		86,550
5-6	Mason, inv. 881	4,007		90,557
5-6	Adams, inv. 882	1,000		91,557
5-7	Adams, inv. 883	500		92,057
5-8	Marks, paid		12,000	80,057
5-8	Adams, inv. 884	600		80,657

It's easy to prove the math in the ledger, and you should do it regularly. You would start with the April 30 balance forward, then add the sum of the charges and subtract the sum of payments:

Balance forward	$74,655.00
plus: charges	+ 30,202.00
less: payments	– 24,200.00
Ending balance	= $80,657.00

Now you've verified that the math on the ledger is accurate. And the total is correct as long as the total charges and payments balance to the control card. Remember, the entries to the control come from the source documents. If you've also posted each individual customer's account record accurately, the total of those balances will also equal this amount.

The next step is to post the entries to each account:

Adams				
Date	**Invoice**	**Charges**	**Payments**	**Balance**
	Balance forward			11,350
5-4			11,350	
5-6	882	1,000		
5-7	883	500		
5-8	884	600		2,100

Carson				
Date	**Invoice**	**Charges**	**Payments**	**Balance**
	Balance forward			3,150

Handley				
Date	**Invoice**	**Charges**	**Payments**	**Balance**
	Balance forward			34,305

Marks				
Date	**Invoice**	**Charges**	**Payments**	**Balance**
	Balance forward			12,000
5-5	879	15,000		
5-8	ck #		12,000	15,000

Mason				
Date	Invoice	Charges	Payments	Balance
5-6	881	4,007		4,007

Smith				
Date	Invoice	Charges	Payments	Balance
	Balance forward			500
5-4	878	1,095		
5-5	ck #		500	1,095

Thomas				
Date	Invoice	Charges	Payments	Balance
	Balance forward			13,000

West				
Date	Invoice	Charges	Payments	Balance
5-5	880	8,000		8,000

Note that some customer accounts are included in the example which have no entries posted that week. Those accounts have an open balance from a previous period that must be included whenever you check the account file for accuracy.

If you add all of the ending balances, the total equals the last total on the accounts receivable ledger and the control. Check your balances frequently until you become experienced working with a subsidiary system.

Remember that *all* customer records must be included when you check the customer account total. Even those accounts with no current activity must be added in for the record total to match the ledger total.

Balancing the Subsidiary System

To prove posting accuracy, you summarize the totals from your source documents on a control card. At the end of the month, the control will show the balance forward, the total charges for the month, total payments, and the new balance forward. The sum of all customer card balances at the end of the month must always equal the last balance forward on the control card.

Here are the steps you'd take at month end to balance the system:

1) Add all charges entered on account cards, and compare this to the total charges in the month's accounts receivable ledger. These totals must agree.

2) Add all payments posted to account cards. Compare this to the total of all amounts posted in your cash receipts journal for accounts receivable payments. Be sure the totals agree.

3) Bring the balance up to date on each customer account card by adding any new charges and subtracting payments made during the month.

4) Add the account card ending balances. That total must be the same as the control card ending balance, and the ending balance in the accounts receivable ledger.

If you don't balance, check your math on:

- the control card
- the accounts receivable ledger
- the cash receipts journal
- the account cards

If you don't find the error that way, you have to go back to the source documents and verify them against the control totals. That can be a difficult and time-consuming job. So be sure to balance each time you post, before the source documents are filed.

Posting the Journal Entries

Once you've proven that your balances are correct, you post the journal entries for the month from the accounts receivable ledger. No matter how much accounts receivable activity there has been during the month, the totals are posted with two simple journal entries. This keeps the general ledger brief and simple. And all the support you will need is in the subsidiary account.

If you have a lot of trouble keeping your accounts receivable subsidiary records in balance, take these steps to correct the situation:

1) Be more careful when posting, and double-check your math.

2) Consider changing to a write-once or a computerized system. I'll have more to say on this at the end of the chapter.

3) Check the balances more frequently than once a month. Then you won't be under so much pressure at month end when it's time to send out statements. If there's a mistake, you won't have to check the entire month to find it.

Invoices and Statements

Invoice forms vary from one business to another. You might design a form just for your own needs, or use one of many stock forms on the market. Whether you buy them or design your own, invoices should contain:

1) A printed number for you and your customers to use for reference.

2) Terms of payment. If you allow a discount, the invoice should say so. Also show when you expect payment. Some typical invoice terms are net 10 days, the 10th of the following month, or 30 days from billing date.

3) Room for a complete description of merchandise sold or services given. Whether you provide labor, parts, charge for freight, or bill on a percentage of completion basis, your form should fit your unique requirements.

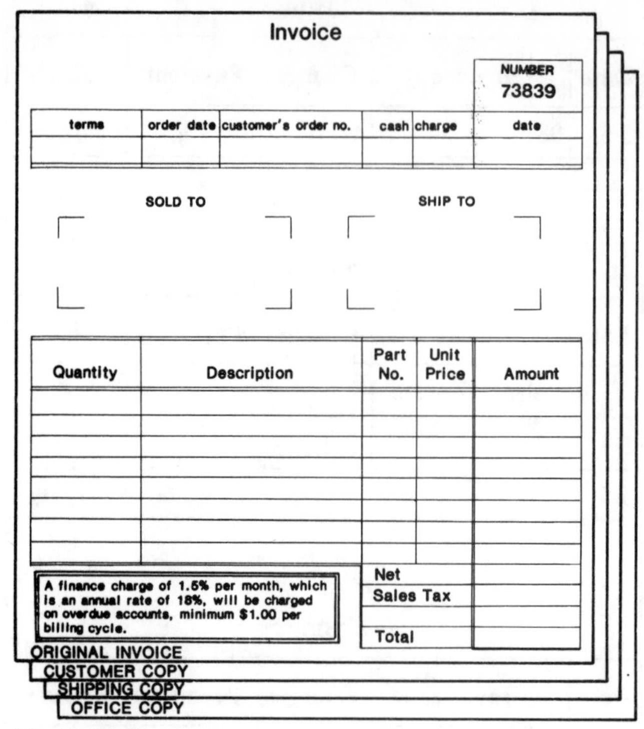

Invoice

NUMBER 73839

terms	order date	customer's order no.	cash	charge	date

SOLD TO

SHIP TO

Quantity	Description	Part No.	Unit Price	Amount
			Net	
			Sales Tax	
			Total	

A finance charge of 1.5% per month, which is an annual rate of 18%, will be charged on overdue accounts, minimum $1.00 per billing cycle.

ORIGINAL INVOICE

CUSTOMER COPY

SHIPPING COPY

OFFICE COPY

Figure 11-6

4) Notice of finance charges. You must let a customer know the annual percentage rate, any minimum charge, and if you assess a finance charge for late payment.

5) Adequate copies. The least you'll need is one for the customer and one for your permanent records. If you ship materials, you'll need a shipping copy. And you might need a copy for posting job cost records. If you want your customers to return a copy with their payments, you'll need a remittance copy also. You can buy or print invoices in carbon sets or with treated paper that creates copies without carbon paper.

Figure 11-6 shows an invoice form suitable for billing parts and materials. The customer receives the original and a copy for their files. The other copies are labeled for their intended use. If your billings include itemized labor charges, you should use a form that suits this better. It might be necessary to custom design your invoices. You could leave the description part of the invoice blank so that one form serves a variety of needs.

Statement

Date

Date	Explanation	Charges	Payments	Balance Due

Figure 11-7

Some customers pay only by invoice, others by statement (supported by invoices). A statement is a good reminder to customers who may have set your invoice aside. A statement is necessary if you have customers that you bill more than once each month.

You'd prepare statements from the account cards at the end of the month. The statement for each customer shows the balance forward at the beginning of the month and any account activity during the month.

A statement is a very simple form like the one in Figure 11-7. The format is the same as the account record.

income journal

DATE	EXPLANATION	DEBIT	CREDIT
	-1-		
5-31	Accounts Receivable	30,202.00	
	Income		30,202.00
	to record charges		
	for May		
	-2-		
5-31	Cash	24,200.00	
	Accounts Receivable		24,200.00
	to record payments		
	for May		

Figure 11-8

Posting to the General Ledger

Once you've balanced the subsidiary system, you're ready to make two summary journal entries. One of them records the current month's charges and the other one the payments. Figure 11-8 shows the journal entries for accounts receivable.

All of the accounts receivable details are contained in the subsidiary. Two journal entries are the only ones made to the general ledger. If anyone needs to check details, they can look in the accounts receivable ledger or the customer account.

Collections

Another important benefit of the subsidiary accounts receivable system is its value in collections control. You always need to know the account status of everyone you've extended credit to. You won't always be able to see this at a

aging list

CUSTOMER	TOTAL	0–30	31–60	61–90	OVER
Adams	2,100.00	2,100.00			
Carson	3,150.00	3,150.00			
Handley	34,305.00		22,100.00	4105.00	8,100.00
Marks	15,000.00	15,000.00			
Mason	4007.00	4007.00			
Smith	1,095.00	1,095.00			13,000.00
Thomas	13,000.00				
West	8,000.00	8,000.00			
Total	80,657.00	33,352.00	22,100.00	4,105.00	21,100.00
%	100%	41%	28%	5%	26%

Figure 11-9

glance by looking at the account cards, especially if outstanding balances go back several months.

The customer account card is a perpetual record, and a customer who has a large volume of transactions with you will have a record going back several months. Try to isolate each customer's record on one account card.

Most of your business may involve one-time charges, with payments received within 15 to 30 days. But you'll want to keep an eye on those repeat customers who may have outstanding balances from one month to the next. The account card lets you see the account history for the entire time you've done business with that customer.

You can see account status with the monthly aging list. The aged accounts receivable report shows the total amount outstanding for each customer. That total is broken down in 30-day periods by invoice date to show how long past due each customer has gone.

The longer you allow a customer's balance to remain unpaid, the more likely you'll never get your money. Bottom-line profits are meaningless unless you also collect the cash. Every month, prepare an aging list like the one in Figure 11-9.

In the example, 41% of all accounts are current (due in 0 to 30 days, or one month); 28% are 31 to 60 days past due (2 months); 5% are past due 90 days; and 26% are past due by three months or more.

You can pinpoint problem accounts by comparing aging lists from one month to another. You don't want to extend more credit to any customer who's already overdue in payment. You can also watch that too much of your total accounts receivable doesn't shift into the past due columns.

You can find out the average number of days it takes to collect money due you. Then you can use those averages to judge individual accounts. You can control your accounts receivable condition to suit your cash flow needs.

Be sure to limit or even stop extending credit when a customer's balance extends beyond the normal collection period.

Here's one case where information from the bookkeeping records is used to create a simple but essential management tool. A business owner needs current, dependable information. The bookkeeping system should be more than an organized collection of numbers. Don't just balance the books and leave them on a shelf. Use them to stay in control of the business, increase profits and avoid losses.

Alternative Systems

If your transaction volume is large, you'll eventually want a system more efficient than hand posting. Posting first to a ledger or log, then to account cards, and finally summarizing on monthly statements, involves a lot of repetitive and time-consuming work.

When you do upgrade your system, be sure you understand the complete manual process first. You'll be better able to manage a more complex system and to keep it running efficiently.

When your accounts receivable posting becomes too time-consuming, you'll probably want to consider one of the following:

1) Write-once systems let you make a single entry to an account card which is recorded on a ledger underneath it at the same time. Posting time is greatly reduced and you cut down the chance for errors.

A write-once system will save you a lot of time. But be sure to develop your control independently and use it to check the details you post to the subsidiary records. Otherwise, any errors will show up both in your accounts receivable ledger and the account cards.

There are a number of good write-once systems on the market especially designed for accounts receivable. If your transaction volume is too large for hand posting, but too small to justify a computer, look into the write-once alternative.

2) Computerized billing saves time. One entry should let you post everything from the invoice to the general ledger. You may not be able to justify automation for your company just for accounts receivable posting unless you have a very large number of transactions. Look for other computer applications like inventory control or estimating to increase your efficiency.

Be prepared for a substantial financial investment for hardware and programs. And be ready to invest some time in learning to use it all. It's not always as easy as the computer or software salesperson makes it sound.

3) Many service bureaus offer accounts receivable services. Some of these are geared toward serving the medical and legal professions. They're best for clients with many customers, each with a limited number of transactions per month. Contractors specializing in residential repair and maintenance services are the most likely candidates for this type of bureau. Contractors are more likely to have only a few customers at any one time, each one accounting for a few large charges and payments.

If the other alternatives aren't acceptable, an independent batch processing service might be the best solution for you.

Self Test

1. *A subsidiary ledger is best used when:*

a) you want to avoid having to post entries twice.

b) there's no more room in your general ledger for more accounts.

c) your volume is too great for efficient management of customer accounts in the primary books.

d) all of the above.

2. *You need to keep track of accounts receivable in order to:*

a) isolate details from the more summarized general ledger.

b) track activity for each customer and prepare monthly billings.

c) make sure that outstanding accounts don't remain unpaid for too long.

d) all of the above.

3. *Individual customer account cards are set up:*

a) for every account.

b) only for customers with an unusually high number of monthly transactions.

c) for those accounts that are past due.

d) none of the above.

4. *You should post accounts receivable subsidiary records:*

a) only after closing and balancing the general ledger each month.

b) just before preparing statements.

c) as often as needed to maintain efficiency.

d) after monthly statements are sent out.

5. *The subsidiary accounts must be balanced:*

a) every week, without fail.

b) before statements are prepared, so that they will be correct.

c) after statements are prepared, because billing customers is a more urgent priority.

d) only when the books have been closed.

6. *All posting should be done from:*

a) invoices or statements, or both.

b) purchase orders received from your customers.

c) the general ledger.

d) the same source each time, in the interest of consistency.

Chapter 12

RECORDS FOR PAYABLES AND PURCHASES

Your accounts payable and purchase records help you plan cash flow. You'll know what payments are coming due over the next month. And you'll avoid late payments that damage your reputation with creditors and suppliers.

You can check your accounts payable commitment before you plan material and supply purchases. You can arrange those purchases to coincide with scheduling demands and available cash. If your records are in order, it's easy to monitor direct costs.

In this chapter, you'll see how to set up a simple but effective procedure for accounts payable where the number of transactions is limited. If your company grows to have a large volume of payables activity, you'll need to formalize and expand on these procedures.

The basic principles and procedures are the same no matter what the transaction level. But the methods change. You may choose to streamline posting by using a write-once system. And at some point a manual system can become so time-consuming that you can justify a computer.

The File System

If you work with only 10 or so suppliers and get no more than 25 bills per month, you can probably handle accounts payable with a simple filing system.

You can use a date-order expanding file to keep track of unpaid bills. As invoices and statements come in, you put them into the appropriate file slot. The slots are numbered 1 to 30 for days of the month. Each slot contains the bills payable that day.

Each day you'd take out the bills due that date, get the proper approvals for payment, and then pay them. If you decide to delay payment, the bill is put back into the payables file under the future date.

The Accounts Payable Worksheet

You should go through the file and list all the invoices periodically — at least once a month. That way you can keep track of what you owe, and to whom. Show the amounts due by due date. This report is similar to the aged accounts receivable list discussed in Chapter 11.

Here's a sample listing of invoices due at the end of a month:

Accounts Payable
May 31, 19xx

Vendor	Total	Due Date		
		6/10	6/15-20	6/21-30
Adams Lumber	416	416		
Carter Supply	845	315	530	
Jamison and Sons	6360	2107		4253
Brown Lumber	1639	1639		
Tolson Materials Co.	305			305
Williams Co.	994		994	
Harper Corp.	500	150	350	
Vickers Tool	2573	2573		
Dodgson Plumbing	494	494		
Bigelow Supply	1844	1844		
Total	15970	9538	1874	4558

When the volume of transactions is large, you'll need a more formal procedure. You'll need a specialized journal and a subsidiary system similar to that for accounts receivable. A separate accounts payable checking account might even be practical.

The following table shows one accounts payable account card used by a residential plumbing contractor who makes several purchases each month from a number of suppliers. His subsidiary system lets him keep track of what he owes and when it's due.

Carter Supply
5123 Industrial Blvd.
Hemiosa, CA 94033

Date	Invoice	Amount	Credits	Paid on Account	Balance
5/10	5-894	315.00			
5/15	5-903	530.00			845.00
6/10				315.00	
6/10	6-143	636.90			
6/15				530.00	636.90

You might consider a separate checking account if:

- You have a large volume of accounts payable due the first part of every month.
- All the accounts are for merchandise purchased during the previous month.

This way you can avoid the monthly accumulation of payables and the related journal. You'll have only one journal entry supported by the total of checks written from the accounts payable checking account.

If you only write 20 or so checks a month, you won't need a separate checking account. But keep this idea in mind if your volume goes up and you're not yet ready to automate your bookkeeping.

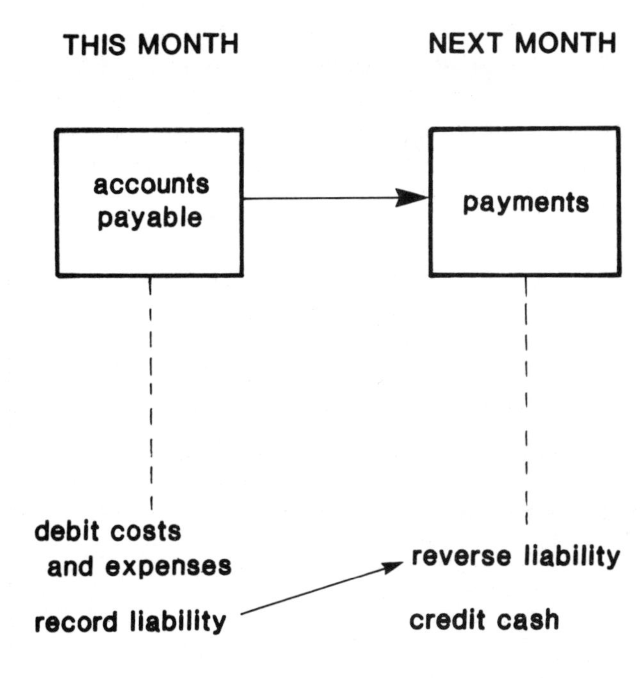

Figure 12-1

Using the Accrual System

Let's review our discussion of accrual accounting from Chapter 2. As you recall, I prefer the accrual system because it produces more accurate financial statements. The statements reflect money owed by and to your company. You can still use the cash method for paying taxes.

You post the amounts you owe into the books through an accrual journal entry. You show an expense for a purchase, even though you haven't paid for it yet.

You reverse accrual entries one of two ways. Either you debit the accounts payable account when the bill is paid, or you make a reversing entry to begin the month following the accrual.

Figure 12-1 illustrates the steps in the accrual system when you reverse the liability at the time you pay the bill.

As an alternative, you can omit the entries to record and reverse accruals, and still have accurate financial statements. You'd just keep all the unpaid bills in a file as described above, and then list them on a worksheet when it's time to include them in financial statements.

This method is completely acceptable, assuming you can keep track of how much you owe and pay your bills on time. There are some problems that can come up though:

- It's easy to misplace payables if they're not formally entered into the books.
- It's hard to go back and draw a financial statement for an earlier period. You may have paid some bills and added others, so you'll have to reconstruct how much you actually owed at the statement date.
- Informal systems are of little value in case of an audit, or if you want to use past performance to make forecasts.

Suppose you want to prepare a budget for the coming year. You'd have only payment records to give you cost and expense facts. Your estimate of future commitments would be distorted because bills are usually paid somewhat later than materials are received and used.

An informal system would be of little use in estimating unless you have a cost accounting system apart from accounts payable. It would be hard to tell which costs related to a particular job.

Make sure your accounts receivable and purchasing records give you the control and the information you need. But don't make them more complicated than they need to be.

Self Test

1. ***The single file system for accounts payable involves placing all outstanding bills in:***
   a) an alphabetical file.
   b) files arranged by invoice date.
   c) files arranged by due date.
   d) none of the above.

2. ***The file system is efficient because:***
   a) the owner can see at a glance what is due and when it must be paid.
   b) that's all you usually need for a limited volume of monthly bills.
   c) it helps avoid setting up another formal record.
   d) all of the above.

3. ***The accrual system makes sense because:***
   a) that's the way professional bookkeepers maintain records.
   b) it's easier than cash accounting.
   c) it's more accurate than cash accounting.
   d) all of the above.

4. ***The best way to manage accrual journals is to:***
   a) make them only for major costs and expenses.
   b) reverse the entry as each payment is made.
   c) Make one reversal entry at the beginning of each month.
   d) none of the above.

Chapter 13

RECORDS FOR PETTY CASH

You'll remember that the most important part of the bookkeeper's job is to be sure there's a complete record of every transaction. You have to be able to trace those records through the general ledger, the journals, and back to the source documents.

To do this, your entries and methods need to be uniform. You post the journals the same way every month. You file your source documents so you can find them easily. You treat every entry the same way.

An efficient bookkeeping system uses the company checkbook as the source document for payments. An auditor can find all the disbursement records in one place. But what about the cash payments you make like the ones in the following examples?

1) You pay a bridge toll each day on your way to and from a job site.

2) You reimburse your job foreman each day for morning coffee and donuts.

3) You buy postage or pay parcel shipping charges.

4) You make a cash contribution to someone who visits your office.

5) You buy a few office supplies at the local stationery store.

Now, how do you treat these cash payments to be sure they're captured correctly in your bookkeeping system?

Some Cash Handling Options

You could write a check for every purchase or expense, no matter how small. But that would be inefficient and impractical. Suppose an employee turns in a receipt for 95 cents. It would cost you more to process the check than the amount you wrote it for. It would also be a nuisance for the employee. And the bookkeeper would have lots of little checks to reconcile at the end of the month.

Another method would be to make a journal entry each time you spend cash. That increases the number of entries to your journal, at the same time raising your chances for making mistakes.

And where does the cash come from in the first place? You write a check to "cash" and keep that cash on hand somewhere. You spend the money, and when it's all gone you write another check. At the end of the month you would gather up all the receipts, code them to the proper accounts and post them.

This method might present a problem in reconciling the receipts for cash paid out with the checks you wrote to "cash." There's no apparent connection between the amounts of the checks you write and the journal entries for the receipts.

While this is closer to a formalized petty cash account than the first example, it still lacks control. As you can see in the following illustration, this system is a little sloppy. When you write the check to "cash," the journal entry would look like this:

Date	Description	Debit	Credit
	Miscellaneous expense	10.00	
	Cash		10.00

When you post the coded receipts, the entry would look this way:

Date	Description	Debit	Credit
	Various accounts	7.85	
	Miscellaneous expense		7.85

The difference between debits to the miscellaneous expense account (created by writing the check) and the credits (from the journal entry) should always equal the cash in your desk or file cabinet. It will be hard to find all those entries in the journal and prove the balance at the end of the year. And if you're audited, this is exactly what you're going to have to do.

The Petty Cash Account

A far better way is to set up a petty cash, or "imprest" account. The balance in a petty cash account never changes. It's always the total of cash on hand plus any receipts for which you've paid cash.

Here's how it works. You figure out how much cash you need. Then you write a check to set up your petty cash fund. You pay all cash expenses from this fund.

The fund should include postage even if you have a postage meter. You'll sometimes need stamps for personal mail such as greeting cards to clients or employees. Or, you might allow employees to buy stamps from the fund.

Here's a guideline for figuring how much you need in your fund: You'll want enough to cover your cash and postage expenses for a two- to four-week period. You'll want enough extra so you'll still be able to make change. But you don't want it so large that it's a temptation to your employees.

One company estimated cash expenses at about $30. They used cash mostly for bridge tolls, postage and office supplies. They set up their fund with $50.

Another contractor with many crews in the field had a much higher volume of cash expense. He opened the fund with $200, with about $30 of that in postage stamps.

Don't misuse the petty cash fund. Its purpose is to provide a way to run cash expenses through the books in a consistent way that allows for verification. Don't let your employees use it to cash personal checks. That'll put a drain on your petty cash fund, and you'll end up having to keep money there that should be in the general account.

The Flow of Money

The imprest system is efficient. You write a check for the fund amount from the general account to set it up. The entry consists of a debit to a new cash account, "Petty Cash," and a credit to cash:

Date	Description	Debit	Credit
	Petty Cash	50.00	
	Cash		50.00

petty cash entry

DATE	DESCRIPTION	DEBIT	CREDIT
9-30	office supplies	17.87	
	postage	6.00	
	auto expense	3.00	
	misc. expense	.25	
	cash		27.12

Figure 13-1

The balance in the petty cash account doesn't change unless you decide you need more or less cash to handle your volume of cash payments. You'd change it by writing another check to increase the fund, or by redepositing money from the petty cash fund back into the bank account. The general ledger cash total consists of the balance in the asset account "cash" plus the unchanging balance of petty cash.

The ledger balance is the amount in the petty cash box. The cash box total is the sum of:

1) Cash

2) Stamps

3) Receipts

During the month, you pay out cash in exchange for receipts. You get a receipt for a $1 toll, and you reimburse your employee with a dollar bill. The fund balance doesn't change.

When the fund is depleted of cash, you code all the receipts in the petty cash box by account. Then add them together to see if their total plus the cash and postage remaining in the box equals the amount the fund started with.

If you've made a mistake in making change, you'll be over or short. In that case, you'll have to make a journal adjustment to miscellaneous expense to correct the balance.

You use the combined receipts as documentation to reimburse the fund. If your petty cash fund was originally set up with $50, here's what might happen at the end of the month:

1) You have $22.88 left in the fund that includes $4.50 in stamps.

2) You have $26.87 in receipts. $17.87 is for office supplies, $6.00 is for postage, and 3.00 for auto expenses. The total of cash, postage, and receipts is $49.75, so you're 25 cents short. You'd write a check for $27.12 and post it like Figure 13-1.

Now you file the receipts with the "paid bills" and cash the check to reimburse the petty cash fund. The total of cash and stamps totals $50.00 again.

The general ledger account "petty cash" hasn't changed. The transaction took place through the checkbook. And the payment is completely documented by receipts.

A good way to file petty cash receipts is to staple them to a check request or summary sheet like the one in Figure 13-4, and put them in a special "petty cash" file. Figure 13-2 shows you how the imprest system works.

The Petty Cash Voucher

Sometimes you'll be asked to reimburse expenses when there isn't a receipt available. Maybe an employee lost a parking or toll receipt. Or the owner made a charitable contribution for cookies or a candy bar and didn't get a receipt. In that case you need to create a source document for your file.

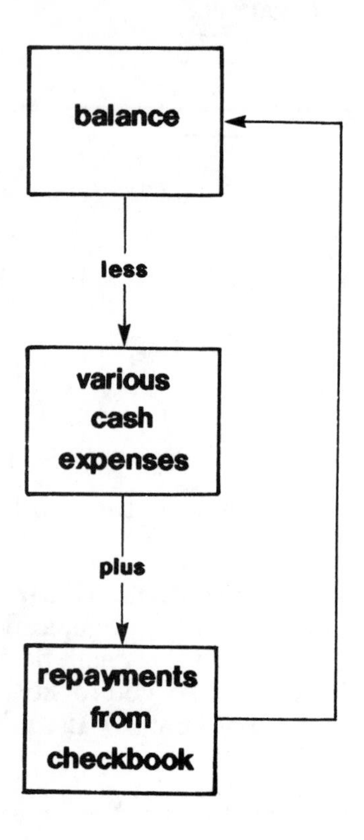

Figure 13-2

You can buy petty cash voucher forms or receipts like the one in Figure 13-3 at stationers and office supply stores. You should use this form for *all* payments from petty cash. Attach the receipts to the back of the form when they're available. There are several reasons to do this.

1) Uniformity. If you use the same form for all cash payments, they'll be easier to recognize, post and file.

2) The voucher/receipt form is an easy-to-read source for the amount and date of reimbursement.

3) You can have the person being reimbursed initial the voucher to show they've received payment. You should also mark the receipt "paid."

4) You can record the account code when you make the payment. This will simplify your job at the end of the month when it's time to post and close the books.

Reconciling the Account

Whenever you replenish the petty cash fund, you need to summarize its activity. For this, use a form that does the following:

1) Breaks down the payment by general ledger account.

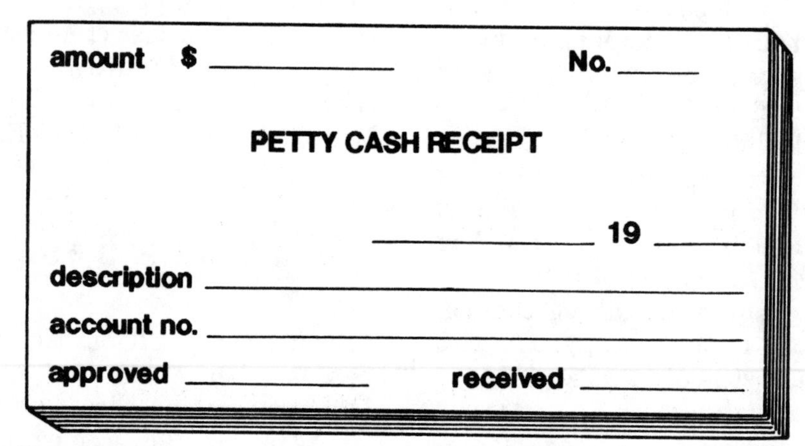
receipt form

amount $ ____________ No. ______

PETTY CASH RECEIPT

____________ 19 ______

description ____________

account no. ____________

approved ____________ received ____________

Figure 13-3

petty cash reconciliation

date 9-30-88

beginning balance 9-15 19 88 $ 50.00

paid from petty cash:

description	amount
office supplies	$ 17.87
postage	6.00
automotive	3.00

total $ 26.87

cash short (over) .25

total paid out $ 27.12

fund balance $ 22.88

paid to petty cash:

check 4316 9-30, 19 88 $ 27.12

ending balance 10-1- 19 88 $ 50.00

Figure 13-4

2) Shows the complete reconciliation of the account for the period.

3) Shows the date and check number for the fund reimbursement.

Figure 13-4 shows a form for reconciling petty cash. It's an efficient form since it both reconciles the fund and summarizes the expense accounts affected by the transactions.

With this system, all cash payments are treated the same way. They all finally go through your checkbook and are under the normal control of your bookkeeping system.

You handle cash the same way as you do bills paid from your checking account. There's no need for a special journal. The check itself is broken down by account categories listed on the reconciliation form.

Securing Petty Cash

As long as you reconcile the petty cash fund each time it's reimbursed, there's usually no need for a special audit. You should have some controls, however.

1) Only one person, usually the bookkeeper, should have access to the fund.

2) Keep the petty cash fund and receipts in a locked box, drawer or file.

3) Have the fund audited periodically by *someone other than* the person who manages it. The auditor should check the paid bills file and verify the reconciliations.

4) Stamp all receipts "paid" when you make reimbursement. You can buy an inexpensive rubber stamp for this in any stationery or variety store. That way, no one can get into the paid bill file and resubmit receipts that were reimbursed earlier.

The petty cash fund is a convenience. It shouldn't take a lot of time and effort. Its purpose is to treat exceptions to the uniform procedure you've set up for documenting, posting and verifying payment entries.

Self Test

1. The purpose of the petty cash fund is to:

a) put cash aside as a reserve for unexpected liabilities.

b) pay only those expenses that come C.O.D.

c) provide consistent documentation for business expenses paid in cash.

d) none of the above.

2. Cash payments:

a) are just as deductible as payments made by check, but must be verified.

b) often are not well documented, because without a petty cash fund, there's no dependable, consistent way to record them.

c) should be handled in a manner consistent with the established bookkeeping procedure.

d) all of the above.

3. An imprest account is one whose balance:

a) never changes in the general ledger.

b) is constantly changing in the ledger due to the varying levels of cash payments.

c) is zero, until a cash expense occurs.

d) doesn't even appear on the books.

4. The petty cash fund may include:

a) currency and coins.

b) postage stamps.

c) receipts for paid bills.

d) all of the above.

5. *When the fund's balance falls:*

a) the reduction is posted to the petty cash account in the general ledger, which acts as a reserve for cash expenses.

b) a check is written to reimburse the fund to its full balance, without altering the general ledger account total.

c) that is a sign that someone is taking money without authorization.

d) all of the above.

6. *The petty cash fund is reconciled by:*

a) an audit of the "paid bills" file.

b) tracking all cash payments during the month.

c) adding cash, postage and receipts on hand.

d) verification by someone other than the employee who handles cash.

Chapter 14

JOB COST RECORDS

The job cost subsidiary system is a critical part of a contractor's bookkeeping procedure. You need current and accurate information to let you know whether each active job is running on schedule and within budget.

Your job cost records have to be easy to read and understand. You must compare them to final job estimates so you'll know whether your pricing and estimating systems are working. In this chapter, I'll show you how to set up and post job cost records so they'll be of most value to your business.

Break Down Costs by Job

As a builder, you'll often have several jobs going at once. You need more than a balanced general ledger to let you know how well your business is doing. You have to watch schedules carefully in order to meet promised completion dates. And you need to monitor job costs constantly. A few days delay or a few hours of extra labor charges can quickly turn a profitable job into a loser.

If your cost accounting is inefficient, or if you don't check the records regularly, you might think everything is all right, when in fact there are problems.

Suppose you complete three jobs one month. The books show you made money. But when you look closely at those three jobs the picture changes. Two of them lost money. Fortunately the third made enough to cover the losses.

When you examine the job cost records, you can see what went wrong with the first two jobs. Knowing the reasons for past losses won't increase past profits, but it could help avoid similar problems in the future. The value of hindsight is limited. There is much more value in an effective job cost procedure where the records are used to monitor a job while it's going on.

There are five broad divisions of job cost information:

1) Payments to subcontractors. When you pay a sub to do part of a job, that's a direct cost. Depending on your company's size

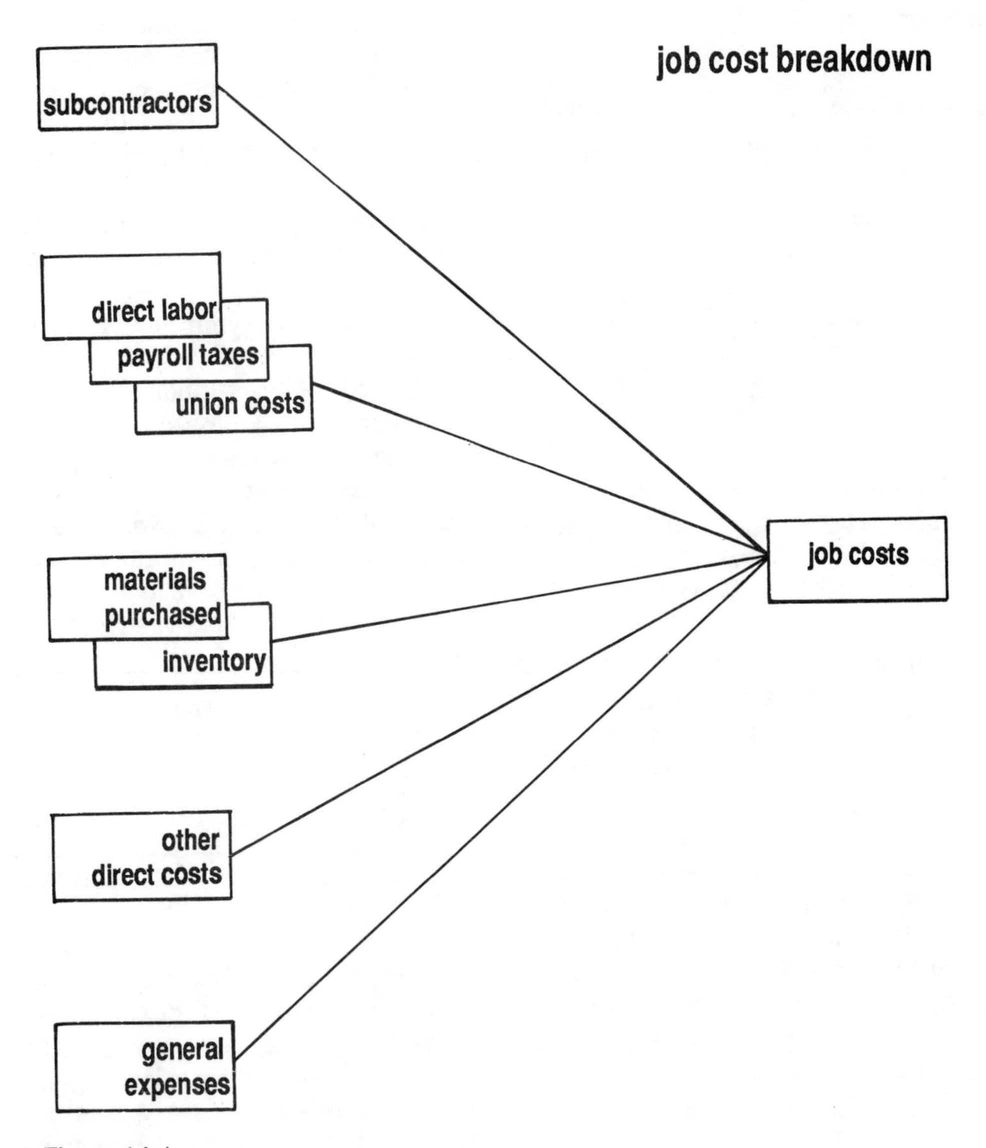

Figure 14-1

and specialty, subcontractor costs can be a major part of the total job. You have to be able to compare your original estimate to the amount actually paid.

2) Direct labor includes the amount you pay to your own crews. In addition to hourly wages you also pay payroll taxes, employee benefits, and sometimes union fees. You need a way to track those expenses and assign them to each job.

3) Materials. If you don't keep inventories on hand, the record is fairly straightforward. Each purchase is applied to a particular job. But if you also remove materials from your own inventory, you need some form of requisition to charge materials by job.

4) Other direct costs. These may be freight charges, payments to designers, architects, engineers, draftsmen, local licensing and bonding agencies.

5) General expenses or overhead. When you estimate a job, you need to add a percentage to cover your general operating expenses. With a tracking system, you'll be able to judge whether your allowance for overhead is enough.

Figure 14-1 illustrates the parts that make up the job cost breakdown.

The Labor Breakdown

The most complex direct cost is the breakdown for direct labor. You have to translate total personnel costs into labor cost per job.

Suppose you have 22 employees working on six different jobs at once. Some may work on just one job, while others may work on all six. Their time is broken down by job number. But to calculate job costs, you can't just multiply the employee's pay by the number of hours they spend on each job. You have to add your payroll costs and employee benefits.

One way to calculate labor costs is to use a yearly average based on total payroll costs for all employees. But there are so many variables in this method that it's an estimate at best, and not very accurate.

A better way would be to calculate total labor cost for each employee using a format like the one in Figure 14-2. You add employer expenses to the gross wages to get the true cost of direct labor.

This gives a more accurate figure, since it takes into account variations between employees. Your contribution for social security will be a lower percentage of total wages for an employee who makes more than the FICA ceiling, for instance. And union fees will vary between trades.

This way you can assign a unique factor to *each* employee, as you'll see in a moment, rather than using an *average* factor for *all* employees.

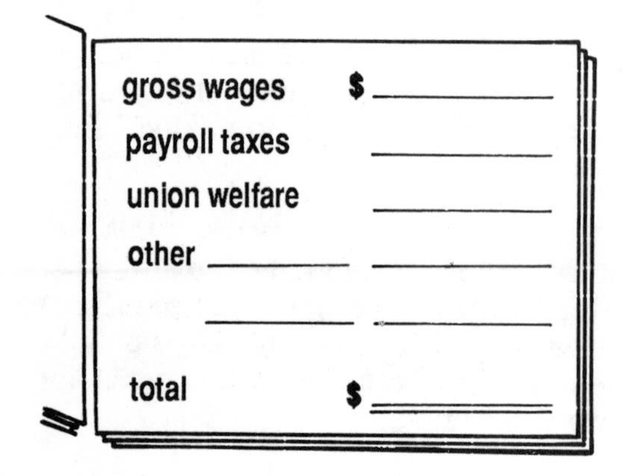

Figure 14-2

If all your employees worked on just one job at a time, posting job cost records for labor would be simple. But in most cases, you put crews together and assign them to various jobs each day. Some crews might even be broken up and reassigned during the day.

You need to design a time card that allows you to break employee hours down by job. Figure 14-3 is an example.

Once you know your hourly labor costs and have a breakdown of each employee's hours, you can figure job direct labor costs.

Suppose your employee, A. Brown, worked on six different jobs one week for a total of 38 hours. Brown's pay rate is $12.35 per hour. When you add employer costs, Brown actually costs you $15.11 per hour. Now you can multiply the hours for each job by your cost to get the proportion of Brown's labor you'll charge to each job.

Job #	Hours	Cost
6	11	$166.21
12	16	241.76
19	4	60.44
4	2	30.22
23	5	75.55
Total	38	$574.18

If you use a payroll service, they'll probably do the breakdown by job for each employee for you. Then you apply an employer cost factor to each employee's gross pay per job to get your job cost figures. Here's how that works using Brown's salary and cost including employer expenses.

First divide the actual employer cost by the employee salary to get the overhead factor.

$$\frac{15.11}{12.35} = 1.2235$$

If another group of employees earn $13.50 per hour, but your total cost totals 17.42, the factor for this group would be:

$$\frac{17.42}{13.50} = 1.2904$$

Now you can prepare a worksheet to add the additional costs to the gross pay for each employee. It would look like Figure 14-4.

Time Card

employee ______________________ week ending __________

	JOB NUMBER								total hours
MON /									
TUE /									
WED /									
THU /									
FRI /									
SAT /									
SUN /									

checked ______________________ approved ______________

date paid __________ check __________

Figure 14-3

		JOB BREAKDOWN				
Employee	**Total**	**6**	**12**	**19**	**4**	**23**
A. Brown	469.30	135.85	197.60	49.40	24.70	61.75
factor	1.2235	1.2235	1.2235	1.2235	1.2235	1.2235
total	574.18	166.21	241.76	60.44	30.22	75.55
T. Smith	540.00	135.00	0	0	175.50	229.50
factor	1.2904	1.2904	1.2904	1.2904	1.2904	1.2904
total	696.82	174.20	0	0	226.47	296.15

Figure 14-4

Direct Labor Allocation

employee	direct labor	JOB NUMBER			
total					

Figure 14-5

Next you can summarize the information for each employee on a form like the one in Figure 14-5. You'll have all the information you need to post the job cost records.

On this worksheet, you list each employee's actual cost in total and by job. When you add the column totals, you'll have each job's total direct labor charge for the pay period.

Materials Allocation

The second major source of direct costs is materials. You can assign each purchase to the proper job when you buy direct and have materials shipped to the job site. Even a single invoice with materials for several jobs can be broken down easily if you use a form like the one in Figure 14-6.

Materials Allocation

supplier	total materials	JOB NUMBER			
total					

Figure 14-6

This is an expansion of the payments journal. Rather than breaking out materials into one category, you spread the distribution into columns for each job.

You can also use a sub-account code number if you have a heavy volume of material purchases. For instance, if material purchases is account number 510, you could code materials purchased for job number 12 to 510.12. You don't have to complicate your payments journal with extra columns for breakdowns by job. Just add one column for the job code.

Total	Materials	Job	Labor	Taxes	Supplies

When you're ready to post job cost records, you summarize the costs for each job and then post to the appropriate job cost card.

To avoid duplication, you may find it practical to post job cost cards using a write-once system. Here's how that works:

- On the left side of the board, you record costs on the accounts payable vendor account card.

- As you enter the information, it's duplicated onto the permanent accounts payable journal beneath the vendor's card. That way, you post cash payments at the same time the vendor's account card is updated.

- On the right side of the board, you post the related job cost record, again onto an individual account card. A copy of that entry also goes through to a permanent record of job cost entries, on the same sheet as the accounts payable journal.

And if your accounting is computerized, you can make one entry that's automatically posted to both the general ledger and the correct job cost account.

If you keep a materials inventory, the task of assigning materials to jobs is more complicated. In that situation, materials assigned to jobs come from two sources. You either order them for the job, or you take them from inventory. You have to account for direct costs in both cases.

Materials ordered for a specific job can be posted directly from the invoice. But you need a way to account for materials removed from inventory. You can use numbered requisition forms and account for each one to be sure you don't miss anything. But this procedure is probably safer:

1) Post all materials to the purchases journal when you order them. Treat all purchases as additions to inventory, even those you know will go directly to one job.

2) Write a requisition for anything taken from inventory. Prepare one also for any materials drop-shipped directly to a job site. This procedure does create more paperwork, but any inconvenience is offset by better control. Your records will be more consistent. And all material usage is accounted for in one way — with a requisition.

3) Post to job cost accounts when the requisitioned material is used. For direct purchases, this will be the ship date. When material is removed from inventory, it will be the date the materials are removed. This way, the job will be charged for materials when they're used, not when they're paid for.

4) Use your requisition and order records to check inventory balances. Start with a physically counted inventory. Add purchases and subtract requisitions. Periodically compare your running inventory totals to the actual numbers on hand. This will prove how accurate your records are, and will show how much control you have over inventory.

Other Direct Costs

It's easy to assign costs for subcontractors, licenses, bonding, freight and other such charges. These expenses relate to a particular job, so they can be posted directly to the job cost record.

The Job Cost Procedure

Design your procedure so you're sure to post *all* necessary information. Your job cost information will usually come from more than one source.

1) You'll post material purchases and other direct costs from the payments journal.

2) Direct labor and related costs will come from the payroll journal.

3) You might use a worksheet to collect some costs like total labor or materials taken from inventory.

Figure 14-7 is an example of a job cost card.

Your job cost procedure should let you see the status of any job on a day-to-day basis. That means you'll have to post cost records daily, even if you only post payments weekly or monthly.

Suppose you bill your customer on the 15th and 30th of each month for work in progress. On the 13th, you order $11,000 worth of material and

Job Cost Record

Job Number __________ Customer ______________________________

Date	Materials	Labor	Sub-Contract	Other Direct	Operating Expenses	Total

Figure 14-7

have it shipped to the job site. You won't have to pay for the materials until September 10.

If you post the job cost record from your payment records, you won't bill the customer for the materials until nearly a month after they were used.

This is where a consistent requisition system pays off. Your job cost records must capture all costs when they're committed to the job, not when they're paid for. Even if you use cash accounting for the rest of your bookkeeping, your job cost records should be on the accrual system.

Be sure your cost cards are up to date with regard to subcontractors. If you pay your subs when you get paid, the bills to your customers must include all subcontractor charges for the period. If a sub finishes a job on the 12th and you're billing your customer on the 15th, be sure the sub's charges are included. Otherwise you might not have the cash available to pay him until after a later billing. Set up your cost records to track commitments, not just payments.

The Job Cost Summary

The best cost procedure is one that allows you to see a daily summary of each job if you need to. Figure 14-8 shows you a form that contains information that will help you see whether each job is running as predicted.

Each job should have a schedule that shows the projected timing and rate of expense for the time work is in progress. The job cost summary shows the percentage of completion and all expenses compared to the schedule. It also shows any variance.

Contractors can make good use of this information. If a job is 5 percent ahead in percentage of completion, that's good. But if material costs are several thousand dollars over their scheduled level for this point in the job, that's not so good. The summary will alert you to either cost or timing problems so you can make corrections before profit or performance suffers.

If you add a percentage to your job cost for general expenses, you can include that figure in your costs at any phase of the job. Suppose you've decided to use 5 percent as your allowance for operating expense. You'd total costs for materials, labor, subcontractors, and other direct costs and then add 5 percent.

Job cost records are only valuable if they're current and accurate. If your procedure is too complicated or time-consuming, it won't reveal useful information when it can do some good.

If you have to keep a large volume of job cost information, you can avoid duplication by using a write-once system. And if you have an automated accounting procedure, be sure it includes the capability for posting by job each time a direct cost is entered into the records.

Be sure you account for direct costs when you make the commitment, not when you pay the bill. That will improve your cash flow. You'll avoid the all-too-common problem of having to pay suppliers before your customer has paid you.

Cost Breakdown

Job ______________________ Date __________

	ACTUAL	SCHEDULE	VARIANCE
percentage of completion	______ %	______ %	______ %
materials	$ ______	$ ______	$ ______
labor	______	______	______
subcontractors	______	______	______
other direct costs	______	______	______
general expenses	______	______	______
total	$ ______	$ ______	$ ______

Figure 14-8

Self Test

1. *Job cost records are essential because they:*

a) are used to prepare future bids.

b) are the basis for payments of material and direct labor.

c) show the breakdown by each job.

d) all of the above.

2. *Labor cost records:*

a) are the easiest to keep, because everything you need is on the time card.

b) must include a factor for costs above the hourly rate.

c) are always broken down by an outside payroll service company.

d) do not apply in most job cost systems.

3. *Labor is broken down:*

a) by adding a factor to the hourly rate for additional costs.

b) by calculating the actual total hourly cost.

c) by hour per job, and then the cost is calculated.

d) any of the above.

4. *Allocating materials:*

a) is always easy, because each purchase is identified by job.

b) is complicated when you purchase directly and also use your own inventory.

c) isn't necessary, because a factor is built into the original bid.

d) none of the above.

5. *Direct cost posting comes from:*

a) payroll journals.

b) the payroll account.

c) worksheets.

d) all of the above.

6. *An efficient job cost system is one that:*

a) does not take a lot of your time.

b) provides the contractor with useful information.

c) is kept up to date.

d) all of the above.

Chapter 15

KEEPING THE CHECKBOOK IN BALANCE

As part of the books and records the bookkeeper maintains, the checkbook is among the most important. Checking account records do two things for your business. They keep track of your daily cash condition. And they give you a method to prove the accuracy of your records and your bank's.

Once a month you'll get a statement from the bank that shows your checking account activity. You need to balance that statement against your records. The closing balance on the bank statement usually won't be the same as your checkbook balance for the same date because of timing differences.

Timing differences occur because your bank doesn't record deposits and checks at the same time you do. When you write a check to a vendor you enter it into your check register and subtract it from your balance. The vendor deposits it to his bank account. His bank then presents it to your bank for payment. Only then does your bank know about the check. Often several days pass between the time you write the check and the time it's deducted from your account balance.

In this chapter I'll show you how to keep an accurate checkbook balance. You'll also see that you can reconcile the checking account quickly and with a minimum of frustration. You'll learn how to:

1) Keep an accurate daily balance, in which math errors are eliminated by double-checking.

2) Understand the different adjustments you'll need to make, and how to account for them.

3) Correct errors, either yours or the bank's.

The Daily Balance

Your daily cash balance has to be accurate and error free for two reasons:

1) To anticipate upcoming cash needs. You want to know exactly how much the cash level increases or decreases each day.

2) To make the month-end reconciliation as easy as possible.

A typical business checking account has three checks per page. On the far left is the check stub where you write the date, the payee, and a description of the payment. There's a column for keeping a running total of your balance.

You enter the balance forward from the previous page at the top of the column. Then you add deposits for the day and subtract the checks as you write them.

When you reconcile the bank statement to your balance at the end of the month, you'll have to account for a number of timing differences and adjustments. Before you begin that step, be sure the math is correct in the checkbook. You can do that day by day if you follow these steps.

1) Be sure that the ending balance on each page is carried forward correctly.

2) Begin with the last balance in the checkbook. *Add* the checks you've written that day to the balance. Then *subtract* the day's deposits. The answer should be the same as your beginning balance for the day.

3) Correct any mistakes before you enter the next day's transactions.

You might use a different system such as a voucher check or a pegboard system. In that case, you'll use a worksheet like the one in Figure 15-1 to keep track of your balance.

You enter the beginning balance at the first of the month. Each day you enter the deposits and the total checks written that day. Then you compute the new balance. You can also use the worksheet to record any adjustments or corrections during the month.

Here's an example of a daily balance worksheet. You've recorded transactions for the first three days of the month. The bank statement has arrived, and there's a difference of $178 between the bank's balance and yours. Let's assume there was a returned check for $150, a $13 charge for printed checks and a $6 bank charge. You also discover you've made a $9 transposition in entering a check.

Date	Deposits	Checks	Balance
Forward			4,218.43
1		1,950.00	2,268.43
2	6,341.90	3,246.63	5,363.70
3	500.00	1,003.75	4,859.95
Adjust	(150.00)	19.00	4,690.95
Correct		9.00	4,681.95

Notice that the $150 for the returned check is shown as a reduction to the deposits column. You keep any adjustments to deposits or checks in their respective column. That makes the month-end reconciliation easier.

The Nature of Adjustments

The reason for the monthly reconciliation is to account for the difference between what the bank says you have in your account and what your checkbook shows. Any difference will be due to timing differences and required adjustments.

Timing Differences

These will be any transactions that are "in transit" on the statement closing date. You may have made deposits near the end of the statement period that don't appear on the statement. And some checks you wrote the last few days of the period don't appear either. They'll be recorded in your check register, but not on the bank statement.

For your balance to agree with the bank's, you have to add the deposits to their closing balance and subtract the checks.

Daily Balance

Month ______________

Date	check numbers	deposits	checks	balance

Figure 15-1

Bank Charges

The bank will adjust your balance for a number of reasons, including:

1) Monthly service and activity charges. These could include charges for honoring or returning overdrafts. You have to reduce your checkbook balance to account for these.

2) Returned checks. You must reduce your deposit total when a check you deposited isn't honored.

3) Automatic withdrawals or transfers of funds. You may arrange with your bank to deduct loan payments, for instance, from your checking account. You might also have money transferred automatically to another account such as a payroll or savings account. These have to appear in your

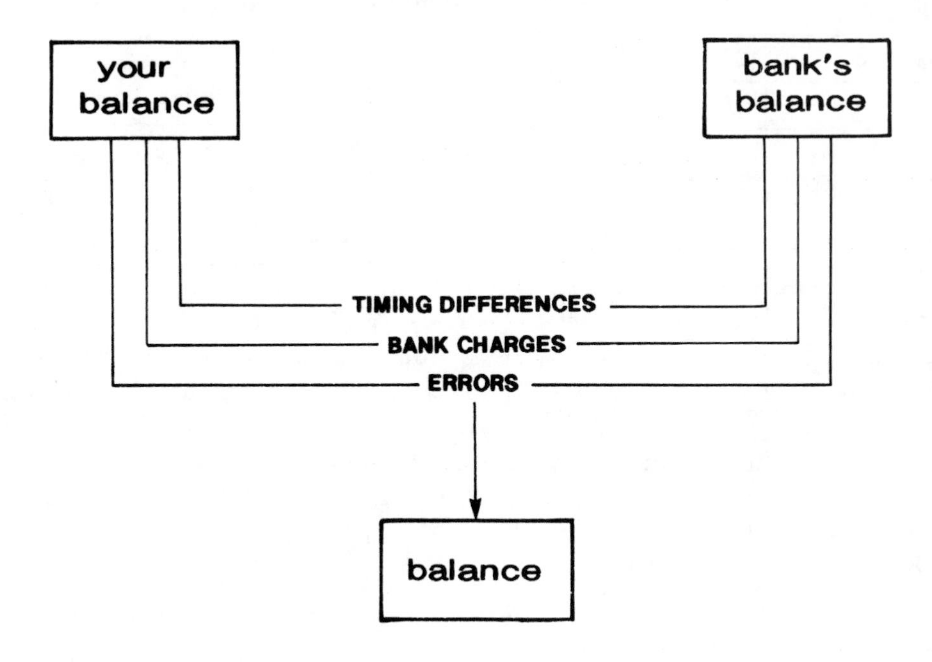

Figure 15-2

your records before you'll agree with the bank statement.

Figure 15-2 summarizes the three things that might cause a difference between the bank's balance and your checkbook.

Adjusting for Errors

If you've made mistakes during the month that haven't been detected, you'll have to correct them after the statement comes. These errors can be simply mistakes in addition or subtraction. You might find these as well:

1) Checks written for one amount but recorded at a different amount. This will require an adjustment to both your checkbook balance *and* to the general ledger.

2) Errors in your deposits. You'll need to adjust your checkbook and your income or accounts receivable accounts.

3) Mistakes that carry back to a previous month.

Taking Action

There might be a variety of checking account adjustments during a single month. Let's discuss the ones you're likely to run into, and what to do about them.

As you saw earlier, the first thing is to enter the adjustments on the daily balance worksheet. Then you make the appropriate journal entry to adjust the general ledger accounts affected by the adjustment. Once you know what the adjustments mean, you'll know how to fix them.

This month you have a $13 charge for printed checks and a $6 service charge. You also discover a $9 error. You recorded an $87 check for materials in your check register incorrectly as $78.

Here's how the adjusting journal entry would look:

Date	Explanation	Debit	Credit
4/4	Office supplies	13.00	
	Bank charges	6.00	
	Cash		19.00
	- To record adjustments -		
4/4	Materials	9.00	
	Cash		9.00
	- To correct math error on check #xxxx -		

You've increased each of the expense accounts and decreased cash for the total. These adjustments have to be accounted for in the books, or they'll come back to haunt you in future months.

Adjustments to Deposits

One area that can cause confusion is returned checks. You receive a check from a customer and deposit it to your account. Then the check is returned due to insufficient funds and charged back against your account. Here's the sequence of events:

1) You receive a $150 check in the mail from customer Green and include it with your daily deposit. You record the deposit in your check register. The bank gives you a receipt and adds the total to your account balance.

2) A few days later, Mr. Green's check bounces. Your bank *reduces* your account balance and sends you a notice of the adjustment, along with Green's check.

3) You call Mr. Green, who tells you the money is now in the account, and that you can redeposit the check.

4) You take the check back to the bank, and this time it clears with no problem.

When these steps occur, you have an original deposit, an adjustment, and a second deposit. You have to make appropriate adjustments in your records to account for all of this.

When your bank returns the check to you, immediately adjust your daily balance register. Then make an adjusting journal entry to correct your general ledger:

Date	Description	Debit	Credit
	Accounts receivable	150.00	
	Cash		150.00
	- Returned check, Green -		

This correctly decreases your cash balance, and also shows an increase in accounts receivable. Note that you also have to show a reduction in cash for the bank's returned check charge, and then increase accounts receivable to recover the charge from Mr. Green.

Date	Description	Debit	Credit
	Accounts receivable	7.50	
	Cash		7.50
	- Bank charge, returned check -		

Now, when Mr. Green replaces the check or tells you it's all right to redeposit the original one, you treat it like a new transaction. You deposit it as usual and make the appropriate entry to the income journal.

Date	Description	Debit	Credit
	Cash	150.00	
	Accounts receivable		150.00
	- Redeposit, Green -		

Even if this happens over two statement periods, your figures should agree with the bank's if you've followed this sequence of entries.

When the Bank Makes an Error

Suppose you made a $500 deposit last month that doesn't appear on your statement. You have a receipt so you can prove you deposited the money. Notify the bank immediately, and supply them with a copy of your deposit receipt. Then follow up until the bank notifies you they've made the correction.

Once in a while, the bank will make a posting error, for instance posting a check twice or processing a check for the wrong amount. Again, you must notify the bank immediately, and furnish copies of the items in question. And be sure you're notified in writing of the correction.

In the case of bank errors, you don't have to adjust your own records. But do be sure to follow through with your bank. The longer you let a problem go, the harder it might be to resolve it.

Accounting for Voided Checks

A voided or canceled check eliminates a check you wrote earlier. If you void a check you wrote this statement period, the adjustment is easy. You just reverse the original entry and it will never become an issue on your bank account. But if you void a check you wrote in a *previous* period, you have to account for it in two places. You need a journal entry, and you also have to adjust your bank reconciliation.

Suppose a trucking company calls to ask you for a $9 payment you sent them two months ago. You realize the check was probably lost in the mail. You issue a stop payment order to the bank and prepare a new check. The first thing you must do is add the $9 to your daily check register.

Then you make a journal entry:

Date	Description	Debit	Credit
	Cash	9.00	
	Delivery expense		9.00
	- Void check # xxxx -		

This entry increases cash and reduces delivery expenses. When you write a replacement check you'll again record the delivery charge and reduce cash. Now you have to adjust your bank reconciliation.

You also have to reduce cash and record a bank charge expense for the cost of the stop payment.

You carry outstanding checks forward from month to month until they clear your bank. Since the above check was canceled, you need to mark it "void" on last month's outstanding checks list so it won't be carried forward.

Preparing for the Reconciliation

Reconciliation is the process of identifying timing differences, adjustments, and errors, and correctly accounting for them. The process will be easy if you watch for these problems:

1) Arithmetic errors in the daily balance worksheet or in the reconciliation itself. Always run a second tape when you make checkbook calculations. You could be out of balance with the bank just because you used the wrong total for outstanding checks.

2) Failure to complete a previous reconciliation or an error in reconciliation. You might fail to post a service charge, or show an increase where a decrease was needed.

3) Incorrect handling of returned checks. Don't forget the reversing journal entries. Otherwise you'll record the same income item twice.

4) Incorrect handling of voided checks. Be sure you remove canceled checks from the outstanding checks list.

5) Remember, you have to begin with a correct and balanced account, or you'll never end up with a balanced account.

Listing Timing Differences

The first step in reconciliation is to list the deposits in transit. These are the deposits you made during the month that don't show on the bank statement. These should be deposits you made right at the end of the period. If a deposit made more than a week before the end of the period is missing from the statement, notify your bank. The deposit might have been posted to the wrong account, lost in the mail, or misplaced.

Use a worksheet like the one in Figure 15-3 to list and total the outstanding deposits. There should only be one or two. Double-check your total.

Next, list outstanding checks, the checks you've written that haven't cleared the bank. Be sure to include both any outstanding checks remaining from the previous statement period, and checks written during the current period that haven't cleared.

Be very thorough in accounting for outstanding checks.

Follow these steps:

1) Some banks produce monthly statements arranged numerically by check number. If yours doesn't, then arrange the checks included with your bank statement in numerical order.

2) Compare the checks paid this period against last month's outstanding check list. Put a checkmark by each check that cleared, or cross out the amount. Be especially careful to compare the check numbers *and* the amount. Use the imprinted amount on the bottom of the check for your comparison. If the bank recorded an amount different from the amount you wrote on the check, you'll discover it by referring to the bank imprint.

Deposits in Transit

Date ____________

date of deposit	amount
total	

Figure 15-3

3) Begin this month's outstanding checks list by copying the numbers and amounts of checks that still remain outstanding from last month's list. Be very careful in transferring these amounts. If you make mistakes, you won't balance. To prove your accuracy, add the unchecked amounts on last month's list, and compare that to a subtotal of the checks you've just written on your new list this month. Use a form like the one in Figure 15-4 to list outstanding checks.

4) Now list this month's outstanding checks. Go through your checkstubs or check register since the last closing date and list all the checks that don't appear on the current bank statement.

Outstanding Checks

Date ____________

check number	amount
total	

Figure 15-4

Adjustments

Date ______________

Description:	ADJUSTMENTS TO: BANK	CHECKBOOK
1. ______________		
______________		
______________	________	________
2. ______________		
______________		
______________	________	________
3. ______________		
______________		
______________	________	________
4. ______________		
______________		
______________	________	________
5. ______________		
______________		
______________	________	________

Figure 15-5

5) Total the checks on your outstanding list. Verify your total by adding it twice. Also recheck the outstanding list for this month against your checkbook.

Listing Adjustments

Adjustments fall into two categories. Those the bank has to fix, and those you must enter into your records. Here are examples of bank errors:

1) Checks recorded for the wrong amount when the imprinted or posted amount is incorrect.

2) Deposits omitted or posted incorrectly.

3) Duplicate entries of checks or deposits.

4) Service charges you don't owe.

These are items you'll have to adjust on your check register and post to the general ledger.

1) Monthly service charges.

2) Fees for stop-payments, returned checks, or transaction volume.

3) Adjustments for returned checks.

4) Adjustments for voided checks

5) Posting errors.

6) Errors from a previous period overlooked or fixed incorrectly.

Enter all of these adjustments on a worksheet like the one in Figure 15-5.

Reconciliation

	BANK	CHECKBOOK
balance		
BANK ADJUSTMENTS:		
plus: deposits in transit		
less: outstanding checks		
other adjustments:		
____________________		
____________________		
____________________		
CHECKBOOK ADJUSTMENTS:		
bank charges		
errors:		
____________________		
____________________		
____________________		
adjusted balance		

Figure 15-6

Describe the adjustment and the method you'll use to make it (such as by journal entry.) Put the amount of the adjustment in the proper column for bank or checkbook changes. If the adjustment is a decrease to cash, put brackets around the amount.

Example: You have two adjustments this month, one to increase cash by $7.00 and another to decrease cash by $15.00. In the "checkbook" column, these should appear as:

a) 7.00
b) (15.00)

Approach the adjustments one at a time. Figure out the *nature* of the adjustment and you'll be able to post it properly. If you've made a mistake in the checkbook balance, decide whether the adjustment increases or decreases your cash balance. Bank charges will decrease cash and be posted to an expense account. Bank credits will increase cash.

locating errors

	BANK		CHECKBOOK	
	DEPOSITS	CHECKS	DEPOSITS	CHECKS
month totals				
plus: deposits in transit				
plus: outstanding checks				
adjustments:				
______				
______				
______				
______				
______				
______				
adjusted totals				

Figure 15-7

The Summary Worksheet

Recheck your worksheets for accuracy. Then use a form like the one in Figure 15-6 to summarize all the timing differences, adjustments and corrections. If you've followed all the above steps correctly, your checkbook balance should be the same as the bank statement balance.

If they don't agree, you'll have to recheck your work. First, narrow down the location for any errors by summarizing the deposit and check entries. Do this both for the bank records and for your checkbook.

The bank usually lists the total of credits (deposits and other additions) and debits (checks and other charges). And if you keep a daily worksheet like the one in Figure 15-1, the totals of each column will give you the corresponding numbers as they appear in your books.

Now summarize the adjustments from the adjustments worksheet, Figure 15-5, and combine those figures with the bank and checkbook totals. Use a worksheet like the one in Figure 15-7 to collect all the figures and pin down the mistakes.

Once you know whether the unreconciled items are in deposits or in checks, it will be easier to find the problems.

The Journal Entry

Remember that identifying adjustments is only part of the reconciliation process. It's not finished until you include the changes into your daily balance record and prepare the appropriate journal entries.

You'll already have made some adjustments when they occurred, rather than waiting for the statement to come. Returned checks or automatic withdrawal notices would be examples of these. Be sure you don't duplicate these adjustments.

All journal entries you make during the month that affect the cash account will have a direct impact on your reconciliation. You must refer to these in order to account for, identify, and correct errors. Those entries will also let you be sure that last month's adjustments were made the way you intended.

Self Test

1. Timing differences are:

a) errors in the bank reconciliation that occur because the bank's closing date is not the last day of the month.

b) errors made in deposit totals.

c) deposits made or checks written that do not show up on the current month's bank statement.

d) any of the above.

2. You should keep track of your daily balance to:

a) find math errors as they occur.

b) report cash availability to the contractor.

c) assist in the reconciliation at the end of the month.

d) all of the above.

3. A bank reconciliation is:

a) accounting for timing differences, adjustments, and errors.

b) making a journal entry to adjust the checkbook total to what the bank says you have.

c) a method for evaluating cash flow.

d) the statement the bank sends you at the end of the month.

4. Errors and adjustments are corrected by:

a) increasing your cash balance.

b) decreasing your cash balance.

c) notifying the bank to fix bank errors you discover.

d) any of the above.

5. The reconciliation is complete when:

a) all deposits in transit and all outstanding checks are verified.

b) the worksheet shows identical balances for the bank and the checkbook.

c) all adjustments are corrected by making journal entries or notifying the bank.

d) you determine that the bank made no errors this month.

6. *When you void a check written in a previous month, you must:*

a) prepare a journal entry to reverse the original check.

b) drop the amount from the outstanding checks list.

c) include the amount as a current month reconciling adjustment.

d) all of the above.

Chapter 16

KEEPING LEDGER ACCOUNTS IN BALANCE

You know you have to balance your cash account to the bank statements each month. What might not be quite so obvious is that you should also check the other general ledger accounts. Even though the books balance, there can be mistakes in assigning accounts. You need to make a special inspection to locate and correct those mistakes.

Let's review for a moment the procedure we discussed in Chapter 7 for closing the books at the end of a period. You carry forward the balances from the asset, liability, and net worth accounts from one year to the next. You transfer the balances from the income, cost and expense accounts to the profit and loss summary. Those accounts begin each year at zero.

The balance sheet accounts are perpetual. You should always be able to explain what their balances represent. And you can check all entries to the current year's income, cost and expense accounts by looking at the source documents.

Create a system for regular monthly review of your general ledger. That way problems won't accumulate. You'll never have to spend a lot of time identifying the contents of any one account.

Identifying Problem Accounts

It's not necessary to review every account each month. You can check some of them quarterly. You may *never* need to analyze others.

- You make only one entry each month to the expense account — rent. If any other entries are made to that account, you'll see the mistake immediately.

- You only write four or five checks a year to the expense account for legal and professional services. A quarterly check of this account is all that's necessary.

- You usually post the accounts for depreciation expense and reserve for depreciation once a year. You only need to analyze these accounts after you close the books each year.

The accounts that you should check regularly are those which:

- Receive many entries each month,

- Are easy to confuse, such as postage and office supplies, or auto expense and repairs and maintenance, or

- Are subject to entries for accrued and deferred income or expenses.

These are the types of accounts that are likely to have errors. Be careful, in your analysis and reconciliation of these accounts, to keep your books as accurate as possible.

Account analysis doesn't require a lot of time. But it's important to be sure you've made all the reversing entries, and have posted entries to the right accounts. That's the only way to be sure that your account balances reflect the true condition of your business.

Reconciling Accounts

Reconciliation is an extra step that's essential to bookkeeping accuracy. Balanced books aren't necessarily error-free. Reconciliation can identify mistakes that would go undetected otherwise. Those errors include posting income or expense items to the wrong accounts, or failing to reverse an accrual.

A mistake in an account that summarizes a subsidiary system such as accounts receivable can have major consequences. You'll think that you have more, or less, than you really have. Any decisions you make based on this false total are going to be bad, and probably expensive. Consider these errors you can find with account reconciliation:

- You make a deposit on an order that will be later divided among several accounts. You post the deposit payment to a suspense account. When the transaction is complete, you'll reverse the entry from the suspense account and charge the amount to the proper accounts. But what if you forget those reversing entries? You'll have an unreconciled balance in your suspense account.

- You may code several purchases to "office supplies" that really belong elsewhere. If you don't assign these expenses properly, you'll have trouble making budget forecasts. The only way to find these mistakes is to trace each transaction for the month back to its source document. If you do find errors, you can correct them with a journal entry.

- The month-end total in your accounts receivable account in the general ledger is different from the total of the customer account cards. You find a journal entry made three months ago to record a bad debt. It was entered into the general ledger, but never posted to the subsidiary system. The balance is still included in the total of the account cards.

- You prepare your quarterly return for payroll taxes. The amount you owe is higher than the amount you show in the general ledger liability account. You discover that you haven't been recording the employer's portion correctly. Again, you can correct the mistake with a journal entry and a change in procedure.

You'll use either a subsidiary record or a worksheet of some kind to reconcile the general ledger. In each example you take a specific action to correct a mistake or resolve a transaction. You reassign the uncleared suspense account balance. You correct the coding errors in the office supplies account. You identify and fix the difference between the subsidiary record and the general ledger for accounts payable. And the error in payroll tax liability calculation points out a problem in your procedures.

Analyzing and reconciling your books makes them more accurate and dependable. It also helps you see where you need to improve procedures. That increases your control over the business.

The Account Summary

Your first reconciliation will probably be the hardest. There may be mistakes that go back months, or even years in the case of balance sheet accounts. Once you have a corrected balance made up of current entries, the reconciliation process is fairly routine. You just identify each month's additions and deletions. Then you post the required reversing or offsetting entries.

If an account is complicated, you can make a summary of its activity for the month. On the first line, list the beginning balance. Follow that with a list of debits and credits, identifying them specifically when necessary. Then write down the account's ending balance. Be sure you make the required correcting or reversing journal entries.

When accounts are posted from two or more sources, divide the account summary into the appropriate groups. For example, you have a liability account called "payroll taxes payable." It's divided into these sub-accounts:

— Federal income taxes withheld

— FICA (withheld plus employer's liability)

— FUTA (employer's liability)

— State disability withheld

— State unemployment insurance (employer)

— State income taxes withheld

These six sub-accounts collectively make up the balance of the liability account. The account balance will be wrong if you make any miscalculations during the month. Errors are more likely because part of the total comes from a calculated journal entry to record employer's liability.

In a case like this, use a form like the one in Figure 16-1 to summarize the account. You can use a column for each type of activity. That way you can isolate the problem areas easily.

The reconciliation procedure is different when you're just identifying what makes up the ending balance. A balance sheet account called "suspense" needs a careful examination each month so it doesn't get out of hand. You use the suspense account to code payments that can't be assigned anywhere else when you make them. The account might contain entries like:

- Deposits received for pending jobs

- Deposits on purchases that will be broken down by account later.

Use a form like the one in Figure 16-2 for the reconciliation when you want to identify the contents of an account at the end of the month. You break down the ending balance by its components and note any adjustments you need to make.

The suspense account can easily accumulate a balance that carries over for many months. It's easy to overlook the reversing journal entry in this type of account. The solution is to create a reminder system when you make the original entry. The closing entry file works well.

The Closing Entry File

Because it's so easy to forget reversing entries, you need a reminder system to be sure that all adjustments are recorded. Here's a closing entry procedure that will make adjusting entries relatively automatic.

Account Summary

Account ______________________________ **Date** __________

					Total
Beginning Balance					
Plus:					
______________					
______________					
______________					
Minus:					
______________					
______________					
______________					
Ending Balance					

Figure 16-1

- Set up a file for each month's closing of the books. Put a written closing procedure in the file. List the posting and balancing steps, the timing of the procedure, cut-off dates, and reconciliation procedures.

- Make a list of the month-end recurring journal entries. For example, if you have a prepaid insurance expense with a 36-month amortization, you must make a journal entry at the end of each month to record 1/36th of the expense. You'd make the same type of entry to amortize organization expenses, record a reserve for bad debts, depreciation, and other recurring entries.

- Also list the correcting or reversing entries you make each month. An example would be a journal entry to reverse the previous month's accounts payable balance.

- Include memos of any entries you made to the suspense file. Resolve them if the information is available to do so. Otherwise, leave the memo in the file for the next closing.

Update your closing entries file each month. Follow your own checklist and procedure and you'll avoid most account reconciliation problems.

Account Worksheet

Account ______________________

Date __________

Ending Balance __________

Description	Balance
Total	

Notes ______________________

Figure 16-2

Figure 16-3 shows some examples of typical correcting entries.

- In Journal 10-17, the $42.00 entry reverses a check coded to the suspense account, properly identifying the payment as Delivery Expense.
- Journal 10-18 corrects a coding error. The expense was actually "printing expense," but was coded to "office supplies."
- Journal 10-19 was a breakdown of sales tax and freight that reduced the book value of a newly purchased long-term asset. In this case, either a payment or a financed balance was coded entirely to the long-term asset account. This entry breaks out the portion of that purchase that goes to current expense.

Your reconciliation procedure does two things. It proves the accuracy of your general ledger balances — and it identifies flaws in your bookkeeping procedures.

When you discover a mistake, correct it immediately. If you let adjusting entries accumulate, the accuracy and timing of your records will be distorted. Make corrections in the current month if possible. When that isn't possible, be sure to do it the following month.

correcting journals

DATE	EXPLANATION	DEBIT	CREDIT
	10-17		
10-31	Delivery Expense	42.00	
	Suspense		42.00
	to record September check #4113,		
	coded in error to Suspense		
	10-18		
10-31	Printing	132.00	
	Office Supplies		132.00
	to correct coding on August check		
	#3816, assigned to Office Supplies		
	in error		
	10-19		
10-31	Taxes, licenses and fees	360.00	
	Freight	114.00	
	Long-Term Assets		474.00
	to correct coding on September		
	purchase of machinery		

Figure 16-3

Self Test

1. Account reconciliation is required when:

a) a large volume of transactions occurs each month in one account.

b) coding errors are frequent, with transactions coded to a particular account each month that belong elsewhere.

c) reversals of entries to the account must be made by way of journal entry.

d) any of the above.

2. *Sources of reconciling entries will include:*

a) original invoices only.

b) subsidiary records and worksheets.

c) historical information.

d) other general ledger accounts.

3. *A suspense account is used to:*

a) record receipts when you think the payee's check might bounce.

b) account for cash overages and shortages.

c) establish reserves for future cash flow.

d) record entries that will be reversed or reassigned in the near future.

4. *Problems reconciling detailed accounts receivable subsidiary records to general ledger entries might indicate that:*

a) entries are not coordinated or controlled between subsidiary and general ledgers.

b) you need to evaluate your posting procedures.

c) a monthly reconciliation is essential.

d) all of the above.

5. *Account summaries and worksheets are used to:*

a) verify transactions and identify components of ending balances.

b) prepare actual amounts posted to the general ledger.

c) prepare recurring journal entries.

d) all of the above.

6. *The closing entry file is used to:*

a) be sure that all recurring journals are made each month.

b) prepare the closing trial balance.

c) create interim financial statements.

d) all of the above.

Chapter 17

CREATING ACCOUNT CODES

The chart of accounts is a numbered list of general ledger accounts. You assign each account a code number. You use the numbers for reference when you post journal entries instead of writing out full account names. If you automate your bookkeeping system, you'll find that computer software works more efficiently with numbered accounts.

Problems and Solutions

One problem with using code abbreviations is the opportunity for error. It's easy to transpose the account number and put the entry into the wrong account. Here's how to avoid this problem:

- Check your posting for correct account assignment as well as accuracy. Be careful of transpositions. Keep your chart of accounts with your checkbook so you can double-check each account assignment.
- Memorize your account codes. Your use of codes will improve with practice, and you'll use some of them frequently enough to make them easy to remember.
- Build control features into your chart of accounts. Use a numbering system that will make it easy to spot errors. For instance, begin all your asset accounts with the number 1. Now if you see a telephone bill coded 175, you'll know there's a mistake.
- In an automated system, you can create more controls by programming restrictions in some accounts. For example, a flexible accounting program can check an account code based on posting limitations you specify. It can require a sub-account within a range of account numbers. And it can limit postings of certain types from certain records. You might specify that entries from the accounts receivable journal will

only be accepted by the cash, accounts receivable, bad debts and income accounts without an error message.

Plan the organization of your chart of accounts carefully. The better you plan it, the easier it will be to use.

Account Logic

Develop your chart of accounts logically. The account numbers should do more than distinguish one account from another. They should make it easier to manage your financial records and prepare reports.

Don't make the numbering system more elaborate than it needs to be. For a small company, a three digit code provides plenty of flexibility. Remember, your general ledger is a summary, so 1,000 distinct account numbers will be more than enough.

There are advantages to limiting the size of your account numbers. It's easier to memorize short numbers. And smaller numbers provide less chance for coding errors and transpositions.

Even with three digit codes, your chart of accounts can be as complex as it needs to be. You can expand to whatever detail you need by using sub-accounts for cost accounting.

To begin, classify your general ledger by major categories according to the way you'll prepare financial statements. You'll have three major groups on the balance sheet side. Those are assets, liabilities and net worth. And for profit and loss accounts, you must classify income, costs, and expenses.

Use the first digit of your account code to identify each of these major categories. The asset accounts begin with 1, the liabilities with 2, and so on. Now you can immediately identify the broad classification for each account number. If the first digit is 1, 2, or 3, it's a balance sheet account. Any number above that is a code for an income statement account.

Figure 17-1 shows how the major categories are defined by the account code's first digit.

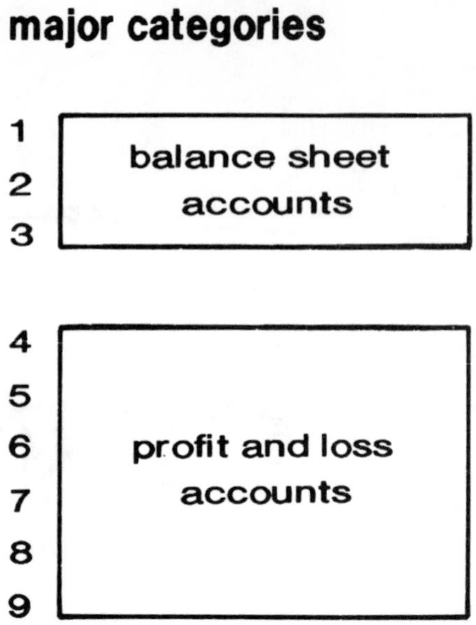

major categories

1, 2, 3	balance sheet accounts
4, 5, 6, 7, 8, 9	profit and loss accounts

Figure 17-1

Major Accounts and Detail Level

Begin with the first digit for major account classifications. Here's a sample list of first digit codes:

Balance Sheet Accounts

1 Assets
2 Liabilities
3 Net Worth

Income Statement Accounts

4 Income
5 Cost of Goods Sold
6 Variable Expenses
7 Fixed Expenses
8 Other Income or Expense
9 Income Tax Provision

Now, expand the numbering system one level. This identifies specific types of accounts within the major categories.

10 Assets

- 11 Current Assets
- 13 Long-term Assets
- 15 Prepaid Assets
- 17 Intangible Assets
- 19 Deferred Charges

20 Liabilities
- 21 Current Liabilities
- 23 Long-term Liabilities
- 29 Deferred Credits

30 Net Worth
- 31 Capital Accounts
- 33 Draw Accounts
- 35 Retained Earnings
- 39 Non-deductible Expenses

40 Income
- 41 Sales
- 49 Returns and Allowances

50 Cost of Goods Sold
- 51 Direct Cost Accounts

60 Variable Expenses

70 Fixed Expenses

80 Other Income or Expense
- 81 Other Income
- 86 Other Expenses

90 Income Tax Provision

Notice how the sub-groups relate to each other. Current assets and current liabilities have similar codes — 11 and 21. Deferred charges have the detail code 19, while deferred credits are code 29.

Listing the Accounts

The next step is to develop the account codes to include all three digits. There will be a code for each account in your general ledger. Notice there are gaps in the numbering system so you can add new accounts in the most logical place.

Following is a sample full chart of accounts.

10 Assets

11 Current Assets
- 110 Petty Cash
- 111 Cash in Bank - General Account
- 112 Cash in Bank - Payroll Account
- 114 Notes Receivable
- 115 Accounts Receivable
- 116 Reserve for Bad Debts
- 117 Retainage
- 118 Inventory
- 119 Investments

13 Long-term Assets
- 131 Land
- 132 Building
- 133 Furniture and Fixtures
- 134 Autos and Trucks
- 135 Equipment and Machinery
- 136 Small Tools
- 139 Reserve for Depreciation

15 Prepaid Assets
- 151 Prepaid Insurance
- 152 Prepaid Interest
- 153 Prepaid Rent

17 Intangible Assets
- 170 Goodwill

19 Deferred Charges
- 190 Deferred Expenses
- 192 Covenants not to Compete
- 195 Deposits
- 197 Organizational Expenses
- 198 Suspense Account
- 199 Accrued Construction in Progress

20 Liabilities

21 Current Liabilities
- 210 Accounts Payable
- 211 Payroll Taxes Payable
- 212 Income Taxes Payable
- 213 Other Taxes Payable
- 216 Accrued Expenses Payable
- 217 Notes Payable, Current Portion
- 219 Deferred Construction in Progress

23 Long-term Liabilities

29 Deferred Credits

30 Net Worth

31 Capital Accounts
- 310 Capital Stock
- 313 Retained Earnings
- 315 Non-deductible Expenses
- 319 Profit and Loss

33 Draw Accounts

35 Retained Earnings

39 Non-deductible Expenses

40 Income
41 Sales
410 Gross Sales
49 Returns and Allowances

50 Cost of Goods Sold
51 Direct Cost Accounts
510 Materials Purchased
514 Direct Labor
515 Subcontractors
518 Freight
519 Other Direct Costs

60 Variable Expenses
610 Travel
611 Entertainment
614 Auto and Truck Expense
616 Repairs and Maintenance
617 Delivery Expense
619 Advertising

70 Fixed Expenses
710 Salaries and Wages
712 Payroll Taxes
715 Employee Benefits
718 Union Welfare
719 Retirement Plans
724 Rent
727 Utilities
730 Telephone
733 Insurance
736 Property Taxes
739 Other Taxes
742 Office Supplies
745 Postage
748 Printing
751 Bonds, Licenses, and Fees
754 Legal
757 Accounting
759 Outside Services
763 Dues and Subscriptions
766 Donations
769 Depreciation Expense
772 Building Maintenance
775 Bad Debts
799 Miscellaneous

80 Other Income or Expense
81 Other Income
811 Interest Income
813 Capital Gains
815 Cash Overages
819 Other Income
86 Other Expenses
861 Interest Expense
863 Casualty Losses
865 Cash Shortages
869 Other Expenses

90 Income Tax Provision
911 Provision for Income Taxes

Adding Sub-accounts

In certain accounts you may want to assign sub-accounts in addition to your three digit code. Payroll taxes payable would be one example. You want to keep your general ledger to a manageable size. But it would be helpful to break out your payroll tax liability among the various taxes you pay.

The solution here is to keep a single liability account, broken into sub-accounts:

211 Payroll Taxes Payable
211.01 Federal Tax Withheld
211.02 FICA Liability
211.03 FUTA Liability
211.04 State Tax Withheld
211.05 State Disability Withheld
211.06 State Unemployment Liability

You can divide other accounts as well. Here are some examples where more detail would be useful:

- To break down reserve for depreciation or depreciation expense by asset.
- To record different sources of income.
- Long-term asset accounts can show detail for each asset in each class.

Don't get carried away, though. When sub-account records become too detailed, you defeat the purpose of the general ledger, which is to summarize information found elsewhere. Don't overdo the sub-accounting procedure. Recognize the point where subsidiary records make more sense.

Here's an example. One contractor first set up sub-accounts for each of his long-term assets. But as his company grew, he owned 18 trucks and a large inventory of other machinery and equipment, office furniture, and small tools. It made more sense to keep records in a special asset inventory log, rather than breaking out each one in the general ledger.

You could use sub-accounts for a very small accounts receivable volume. But if you have more than four or five customers at once, you're ready for a subsidiary accounts receivable system.

Sub-accounts only need to be two digits. That allows you up to 99 divisions for each account. If you ever need to break out more than five or ten sub-accounts, you're better off with a subsidiary system. Just the same, use a two digit sub-account number. One digit would be too restrictive.

The sub-account follows the account number itself and is separated from it by a period:

xxx.xx

Cost Control Information

You can add another level to your chart of accounts for tracking costs. But remember that the reason for account codes is to simplify, not to complicate. Don't add further subdivisions to your account numbers unless you really need them.

You could use a three-digit sub-code for cost accounting. This follows the account and sub-account, again separated by a period:

xxx.xx.xxx

If you have only a few jobs going at once, you can use the two-digit sub-account for cost accounting.

The complete account code including a cost record consists of five levels:

X	Major Category
XX	Detail
XXX	Account Number
XXX.XX	Sub-Account
XXX.XX.XXX	Cost Record

Remember, the more complex your account codes, the longer it will take you to code your transactions. And the easier it will be to make mistakes. Here's a summary of guidelines to use when you develop your own chart of accounts:

- Develop your account codes logically.
- Limit the number of digits in the coding to what you actually need.
- Use sub-accounts when you need just a few general ledger divisions. Use a subsidiary system if you need a more detailed breakdown.
- Use a cost accounting system only for direct costs.

Self Test

1. The chart of accounts is used to:

a) eliminate the need for writing out full account titles each time you prepare a journal entry.

b) manage transactions on an automated bookkeeping and accounting system.

c) set up sub-account and job cost control records for selected accounts.

d) all of the above.

2. *Coding errors can be located by:*

a) running a batch total of the account numbers before you post.

b) using the code and the full account name.

c) carefully reviewing coding, and comparing major account categories to the type of transaction.

d) all of the above.

3. *The first digit of the account code identifies:*

a) whether it's a balance sheet or a profit and loss account.

b) the individual preparing the journal.

c) the customer and job to which the entry is assigned.

d) any of the above.

4. *The first and second digits of the account code identify:*

a) the detail level within categories of the general ledger.

b) the customer and job to which the entry is assigned.

c) the year a job was started.

d) the department to which the transaction is assigned within the firm.

5. *A sub-account is to be used when:*

a) you want to avoid the extra work of setting up a subsidiary record.

b) you run out of room in the chart of accounts.

c) you want to keep track of payments by vendor.

d) none of the above.

6. *A job cost code is to be used:*

a) for every transaction you post.

b) only for direct costs.

c) only for profit and loss accounts.

d) only on a fully automated system.

Chapter 18

STAYING IN CHARGE

Back in Chapter 1 we described the bookkeeping process this way:

- It's a system for recording your financial transactions in journals and ledgers.
- It establishes an "audit trail" that shows where the numbers came from.
- It gives the contractor up-to-date information about cash, income, expenses and profits.
- It collects information for financial statements. These show the profits and the asset value of a business.

Besides these features, your bookkeeping system will also give you planning tools and protection against expensive mistakes or dishonesty.

The Control Feature

To fully protect company assets, the bookkeeping system needs control features built into its procedures. All business managers should look for ways to protect themselves against theft and embezzlement.

These areas require special attention:

- Supervise your inventory. Lock up valuable and small parts. Account for tools at the end of shifts, and take periodic physical counts.
- Design specific cash handling procedures and be sure that all your employees follow them.
- Keep strict control over the checkbook and related journals.

If you think of security controls and bookkeeping as separate and unrelated, neither will be fully effective. The best controls are the ones you include in day-to-day procedures.

Always split cash handling between two or more people. That way, any embezzlement must involve either a conspiracy, or an especially clever scheme. There's no absolute way to prevent a clever thief from stealing money from your business. The best you can do is build controls into your bookkeeping system that make it as difficult and risky as possible.

If yours is like thousands of small companies, you may have only one employee who is your office assistant and bookkeeper. In this case it probably isn't efficient to involve yourself in the routine cash activities. But you *must* be willing to spend a little time overseeing cash and checkbook procedures. Otherwise you're totally dependent on the trustworthiness of your employee. And even this person, who may have been working for you for years, can be sorely tempted when there's a lot of cash flowing through his or her hands. Don't subject your employees to this. Meet your responsibility to supervise, and you'll make it easy to be honest.

Consider this situation: Suppose you have one employee who opens the mail, writes all the checks, and balances the bank account each month. It would be fairly easy for that employee to embezzle and cover their tracks so it would take an in-depth audit to reveal the theft.

Here are three steps that require a very small commitment of time to help you protect yourself:

1) Review the status of accounts receivable regularly. A common embezzlement technique involves writing off an account receivable and keeping the money paid against it. Be sure you review and approve any write-offs. Be personally involved in the collection process so you're sure the customer really hasn't paid.

2) Be sure *you're* the one to open and inspect the bank statement each month. Take a few minutes to look at all the checks. You'll have a chance to review the payees before anyone can remove any checks from the statement. You can also compare any payments you don't immediately recognize to your disbursements journal or the paid bills file.

3) Be sure you're the only one who can sign checks. And no matter how busy you are, don't sign checks without looking them over carefully. Be sure you know where they're going, and for what.

This review should only take 30 minutes or less per month. It's a worthwhile protective step.

Don't use a signature plate in a check imprinter. If your bookkeeper uses a signature plate, then balances the books and reconciles the bank statement without your scrutiny, you've given up all your cash protection.

Take time to review the general ledger from time to time. Pay special attention to these points:

- Make sure the ledger balances. Check entries on the worksheet or trial balance.

- Investigate any unusual or unexplained journal entries. Pay special attention to entries to the cash, accounts receivable, bad debt, and miscellaneous expense accounts.

- Check any balances carried forward in the journals. One method of embezzlement involves carrying the wrong balance forward from the end of one page to the beginning of the next one. If a person does this, the totals still balance, but the wrong amounts are posted to the general ledger. This way a dishonest employee can remove cash without putting the cash account out of balance.

One sadder but wiser contractor uncovered this scheme: An employee, who had access to the daily worksheet, wrote a check for $100 to himself, and forged the owner's signature. He didn't want the check to show in the disbursements journal, so he marked that check number "void." He added the totals for the page, balanced the amounts reported for each column, and cross-footed the totals. Anyone checking that page could verify the "correct" balance.

On the next page of the journal, the employee increased the balance forward for the "check amount" and "materials" columns each by $100. Now the amount of the theft has been entered into the system and anyone who checks the new page can still verify a correct balance. The total disbursements entered in the cash account are accurate. The checking account will balance after the check clears. All the employee has to do is destroy the check made out to himself when the bank statement arrives, continuing the illusion that it was voided.

This builder now monitors his books closely. He opens and reviews the bank statement each month. He also insists that he see all voided checks and that they not be destroyed. He could further protect himself by checking the journal balances forward. And he shouldn't allow the employee who reconciles the bank statement to also be the one to write checks.

Embezzlement and theft of supplies or inventory is an expensive problem. You probably can't eliminate it completely. But you can avoid most cases of loss from employee theft with a few simple controls. Companies that do suffer losses from embezzlement have one attribute in common. There is a lack of awareness or concern on the part of the owner or a responsible manager.

The Management Function

A smart contractor will be actively involved with the financial management of the business. An accountant can be an important partner in that effort. The accountant's specialty is verifying and analyzing the company books, and making projections based on what they contain. *Builder's Guide to Accounting*, also by this author, will show what an accountant can do for your company. It's listed in the order form in the back of this manual.

Successful builders know that planning is a critical function. Their knowledge and experience help them to anticipate the future based on historical information.

Here's the point where the bookkeeper's job ends and the owner's job begins. The bookkeeper's job is to present information that's accurate and revealing. And the owner, alone or with the advice of an accountant, must be able to interpret that information and apply it to future action. Analysis of the books lets you see an emerging problem, anticipate a cash flow crisis, or take steps to avoid a schedule delay.

Don't make your bookkeeper's job difficult. Recognize the value of the information that's in your bookkeeping system. Appreciate your bookkeeper's contribution to your company, but don't expect too much.

A skilled and experienced bookkeeper may be able to do many of the things an accountant is trained to do. But it may be unfair to expect that. It would be unreasonable to turn over the analysis and financial planning responsibilities to an employee who isn't equipped to handle them.

Let your bookkeeper confine his or her activities to keeping the books. Make sure the job description is clearly defined. Consult an *accountant* for help in planning and interpretation. Don't expect executive abilities from a clerical employee, any more than you'd pay a $150-per-hour accountant to manage your petty cash fund or check the math on time cards.

Be sure everyone's roles are plainly defined.

- As the business owner, you make the final choices, especially the tough ones. Financial decisions shouldn't be delegated to the bookkeeper or even the accountant. Depend on those people for the information you need, and then proceed.

- Your accountant is an analyst. He or she specializes in recognizing how today's events will affect the future. The accountant can anticipate and recommend corrections to poor cash flow, out-of-control expenses, inappropriate inventory levels, low profits and too-high tax liabilities.

- The financial information you and your accountant use will come from your books. So your bookkeeper's job is to provide dependable, accurate and timely information.

The functions of the owner, accountant, and bookkeeper might overlap. In some operations, the bookkeeper might prepare budgets or be responsible for other accounting duties. And the owner might balance the bank statement as a cash control security measure. There won't be problems as long as everyone knows what's expected. Overlapping isn't a problem unless it happens without clear definition.

The Automation Alternative

When financial management problems occur, it's usually not because of overlapping responsibilities. The fault is more likely a systems problem. If your bookkeeping system is inadequate for your transaction volume, accuracy, timing, and planning will all be affected.

At some point a growing business will have to consider automating its books in order to keep up. Automating to make your system more efficient is a worthwhile investment.

But don't automate for the wrong reasons. Unfortunately, many small companies have regretted automating their bookkeeping functions. Problems come up when a company expects a computer to "fix" a poorly managed manual system.

Is it time to automate your bookkeeping system? Evaluate your transaction load today compared to a year ago. Are you processing many more checks and receipts? Is it taking you much longer to post and balance the books? If so, automation could save you money.

Recognize a computer's values and limitations. In its best application, a computer is most efficient for processing many transactions with similar attributes. Remember that if you have a complicated manual bookkeeping system, an automated system will be relatively complicated as well.

Many computer software companies offer fine bookkeeping programs. They're designed to handle a wide variety of entries. The best ones prompt you through the entire posting process and cut out nearly all manual calculations. In one segment of a program you can enter cash disbursements and the program creates a journal and automatically posts your ledger. Some programs include accounts receivable and cost accounting components. All of this saves you a lot of work — if you know what you're doing.

That's a key point. You *must* know the bookkeeping system thoroughly before you automate. Otherwise, going from a manual system to an automated one will turn your difficulties into nightmares.

In one company I observed, the builder hired an employee with no bookkeeping experience. Learning on the job, the employee struggled with the books every month. It was a constant battle to keep the ledger in balance, reconcile the bank statements and keep the posting up to date. The owner saw the problem and thought a computer might help.

They bought a computer and took a training course. Then they installed their bookkeeping system. But they were disappointed with the result.

- The system had rules of its own to learn and apply. Posting actually took longer than before.

- Because the bookkeeper hadn't mastered the concepts and procedures of the manual system, mistakes were compounded. They were harder to find, and still harder to correct.

A computer is a tool and not a substitute. If your transaction load has become overwhelming, a computer can save you time. But be sure you're in charge first. Don't automate to solve a lack of competence. For more information about when and when not to get a computer, and how to go about automating your system, order a copy of *Computers: The Builder's New Tool*, also described in the order form in the back of this manual.

The bookkeeping process is satisfying because there's only one absolute, right answer. When you've got it, you know you've got it. As with any occupation, skill comes from study and practice. A beginning bookkeeper complains, "These books just won't balance." Once they gain experience, they learn to admit, "I've done something wrong."

As a builder, you've gone through the same process in your career. At first, the hammer kept hitting your thumb. Eventually, you learned to correct your technique, admitting that it was your own action that caused the pain.

And that's what happens in bookkeeping, too. You may be frustrated in the beginning. But once you've learned the rules, the process is logical, precise, and clear. When you master the techniques, you'll be in control. The books will become *your* books. And they'll be a valuable tool in making your business a success.

Self Test

1. A properly maintained set of books:

a) documents each transaction from the source document, all the way through to the general ledger.

b) includes verified balances for all accounts.

c) is essential for financial statement preparation.

d) all of the above.

2. Control over cash, inventory and other assets:

a) is a specialty for the owner, and should not be mixed up with the routines of bookkeeping.

b) is not completely possible, so that attempts to add control in the books only complicates the job.

c) is an essential and necessary feature of the bookkeeping job.

d) none of the above.

3. Embezzlement is most likely to occur:

a) when one employee handles all cash and bookkeeping functions.

b) when the owner does not oversee the bookkeeper.

c) when the owner is unaware of the need for review and supervision.

d) all of the above.

4. The bookkeeper's responsibilities include:

a) ensuring accurate and complete information in the records of the company.

b) planning, to anticipate future cash flow, tax, or scheduling problems.

c) advising the owner in matters of finance, profits, and cash flow.

d) all of the above.

5. The distinction between bookkeepers and accountants is that:

a) bookkeepers are employees, while accountants are consultants.

b) accountants advise owners and help spot future trends, while bookkeepers provide the information needed to analyze those trends.

c) a distinction of title only.

d) accountants are managers, while bookkeepers are clerical employees.

6. An essential step to take before deciding to automate the books is to:

a) compare prices of hardware and software systems.

b) hire a skilled programmer.

c) evaluate today's transaction load to determine whether automation will be efficient.

d) find the best accounting software on the market.

SELF TEST ANSWERS AND EXPLANATIONS

Chapter 1

Question 1. Answer: a

Explanation: You calculate return on *investment* by dividing the profit by the amount of your investment:

$$\frac{\$5,500}{\$80,000} = 6.9\%$$

Question 2. Answer: c

Explanation: You calculate return on *sales* by dividing profit by total gross sales:

$$\frac{\$5,500}{\$66,000} = 8.3\%$$

Question 3. Answer: b

Explanation: You record all checks in the disbursements journal. Use the general journal to record transactions that don't belong in either the receipts or the disbursements journal. You post the total of all entries of each type in the general ledger.

Question 4. Answer: d

Explanation: Receipts appear in all of these books in one form or another. The general ledger summarizes all activity at the end of the month. The accounts receivable ledger is a subsidiary record organized alphabetically by customer name. And you use the receipts journal to record all earned and paid income.

Question 5. Answer: b

Explanation: The balance sheet is a summary of assets (properties owned), liabilities (debts owed) and net worth (the difference between assets and liabilities).

Question 6. Answer: d

Explanation: Bookkeeping has many benefits for the contractor beyond the routine of posting transactions into accounts. All of the examples are valid benefits to be derived from a bookkeeping system.

Chapter 2

Question 1. Answer: b

Explanation: Cash accounting involves making entries *only* when cash changes hands. It doesn't recognize income in the period earned, or expenses in the period incurred.

Question 2. Answer: b

Explanation: Accrual accounting is a method for recording all income, costs and expenses in the period to which they apply, regardless of when cash changes hands.

Question 3. Answer: c

Explanation: On the cash basis, the company took in $2,900 less than it paid out. Earned income is actually higher, but it won't be booked until payment is received.

Question 4. Answer: a

Explanation: All accrued income and costs or expenses incurred are recognized during the current month. Income of $46,800, less costs and expenses totaling $34,700, net to a profit of $12,100. You recognize the cash receipts and payments in the period when they were earned and incurred, not when the cash actually changed hands.

Question 5. Answer: d

Explanation: Reason out the entry this way: You want to show an increase in the income account, since income was earned during the month. The income account increases as a credit balance, so you have to credit that account. Because you didn't receive any cash, you have to increase the balance in accounts receivable. This is an asset account, and increases with a debit.

Question 6. Answer: b

Explanation: As in question 5, you have to reason out the entry. You receive cash, so you need to debit the asset account, cash. You booked the income during the previous month, when it was earned. You have an accrual in the accounts receivable account, which you must now reverse. That requires a credit to accounts receivable.

Chapter 3

Question 1. Answer: c

Explanation: All types of journals must contain debits and credits of equal value, without exception. This is a basic premise of the double-entry bookkeeping method.

Question 2. Answer: a

Explanation: Cash accounting recognizes income only when cash is received. Earnings are not posted as they are earned under this method.

Question 3. Answer: c

Explanation: The accrual method requires entries for earned income, and for cash receipts as well. Typically, earned income requires a debit to the accounts receivable account and a credit to the income account. And a cash receipt involves a debit to cash and a reversing credit to the accounts receivable account.

Question 4. Answer: d

Explanation: Records aren't isolated in a complete bookkeeping system. Everything is kept in balance through the combination of records. Accounts receivable, cash receipts, earned income, and job cost records each serve a different purpose, but all are affected by journal entries you make in the income and cash accounts.

Question 5. Answer: c

Explanation: Don't expand your bookkeeping system unless you need to. No one system will work for every company. As your transaction volume changes, you must expand or modify your bookkeeping records. Automation and write-once systems are possible ways to handle higher volume, and can save you time and effort. But use them only when they make your system more efficient.

Question 6. Answer: a

Explanation: It makes sense to combine income and cash receipts records in a manual system. This reduces the number of records you have to keep, and still allows you to monitor your accounts and keep them in balance. You also have a summary, all in one place, of all the information that goes into the accounts receivable subsidiary ledger.

Chapter 4

Question 1. Answer: a

Explanation: The accrual system recognizes all income, costs and expenses in the month incurred, even when payments happen earlier or later. You accrue payables to book the cost or expense in the current month, and then reverse them the following month — either by coding payments or with a reversing entry.

Question 2. Answer: d

Explanation: All of the answers are correct. A requisition is a form of internal control. If you use the applicable requisition to confirm payments, you won't make duplicate payments. This way, you can keep track of outstanding payables, and eliminate them once they're paid. In addition, you can use the information on the requisition to identify the job and post job cost records when you place orders for materials.

Question 3. Answer: b

Explanation: By adding every column down, and then adding distribution across and comparing it to the total, you can be sure your journal is accurate before you post the general ledger. This reduces balancing problems you might face later, when you close the books.

Question 4. Answer: d

Explanation: All the answers are correct. Keeping separate accounts doesn't eliminate the need to balance accounts payable, especially when the estimated liability is different from the actual amount paid. You'll have to reconcile two bank accounts every month. And if you make a payment from the wrong account, your books won't be accurate.

Question 5. Answer: c

Explanation: The purpose of an accrual is to record everything in the proper month. You don't post any accounts twice, although the payment of a bill and reversal of the previous month's accrual do offset one another in the same account.

Question 6. Answer: d

Explanation: All of the reasons listed can justify opening additional checking accounts. A high volume of payments may require a separate account. A complete requisitioning system will help accrue costs, and control payments in future months. And specialized accounts can make the reconciliation process easier.

Chapter 5

Question 1. Answer: d

Explanation: You use the general journal to record all transactions that don't fit into the income or payments journals. That includes all expenses paid in cash, errors, adjustments, and non-cash expenses.

Question 2. Answer: b

Explanation: Every accrual must be reversed. If you code accrued expenses as debits to accounts payable the month after an accrual, that will reverse the original credit journal entry. If you prefer, you can make a reversing journal entry, which is exactly opposite the coding of the original accrual.

Question 3. Answer: b

Explanation: The accrual is made so that each month's financial statement will accurately report earned income and incurred costs or expenses.

Question 4. Answer: a

Explanation: All accruals must be reversed. Without the reversal, the current month's financial statements will be inaccurate. For example, you accrue an expense with a journal entry, and then also write a check in payment for that expense. If you don't reverse the accrual, the expense will be doubled in the books.

Question 5. Answer: c

Explanation: You only use adjustment journals to correct errors or to enter bank charges and returned checks.

Question 6. Answer: d

Explanation: Non-cash expenses include amortization, depreciation, and booking of reserves for bad debts. You don't pay these expenses through the checkbook, but they're expenses you need to record for tax purposes.

Chapter 6

Question 1.

The following is a correctly posted summary of the general ledger.

	Debit	Credit
Cash	713.52	
Accounts Receivable	25,901.16	
Inventory	14,300.00	
Fixed Assets	62,814.00	
Reserve for Depreciation		18,211.00
Prepaid Assets	610.40	
Accounts Payable		1,422.93
Payroll Taxes Payable		704.16
Notes Payable		749.05
Net Worth		82,219.91
Income		22,485.10
Materials	9,372.79	
Direct Labor	4,532.00	
Operating Supplies	152.05	
Freight	85.40	
Rent	700.00	
Salaries	2,800.00	
Payroll Taxes	704.16	
Office Supplies	271.25	
Bank Charges	8.00	
Telephone	397.82	
Utilities	85.55	
Repairs	32.00	
Postage	44.00	
Subscriptions	15.40	
Travel	106.73	
Printing	84.77	
Advertising	102.00	
Professional Fees	400.00	
Licenses and Fees	15.00	
Insurance	144.15	
Depreciation	1,400.00	
Total	125,792.15	125,792.15

Chapter 7

Question 1.

Following are the corrected journals and ledgers, assuming the original error in the journal was in the cash credit:

Journal		
Account	Debit	Credit
Materials	4,316.85	
Office Supplies	201.40	
Telephone	100.00	
Cash		4,618.25

Cash		
	Debit	Credit
forward	6,218.00	
		4,618.25
balance	1,599.75	

Office Supplies		
	Debit	Credit
forward	1,150.35	
	201.40	
balance	1,351.75	

Materials		
	Debit	Credit
forward		12,954.16
	4,316.85	
balance		17,271.01

Telephone		
	Debit	Credit
forward	3,812.19	
	100.00	
balance	3,912.19	

Question 2. Answer: d

Explanation: Errors can appear in any of those places. Follow a methodical process to balance the general ledger. Remember that all debit entries minus all credit entries must equal zero.

Question 3. Answer: c

Explanation: When the total of the digits of your out of balance figure add up to 9, chances are you've transposed an entry. In this case 1 + 4 + 4 = 9. Also, if you divide $1,440 by 9, you get 160. This is a clue that you may have entered $160 as $1600. Look for the error in the amounts you posted from the journals to the ledger.

Question 4. Answer: a

Explanation: The trial balance verifies the accuracy of the general ledger before you prepare the financial statements. An experienced bookkeeper can skip this step and prepare statements directly from the ledger. But when you're starting out, the trial balance is a useful step.

Question 5. Answer: c

Explanation: When you prepare interim financial statements during the year, you usually don't want to make adjusting entries in the general ledger itself. During the year, the entries are either estimates or they vary during the year. The closing worksheet is for temporary adjustments only.

Question 6. Answer: c

Explanation: You make the final entry when you close the books at the end of each year. You carry balance sheet account balances forward to the next year. But you close all income accounts — sales, costs and expenses — to zero. Then you post the net total to the profit and loss account which is part of net worth. A profit increases net worth, while a loss decreases it.

Chapter 8

Question 1.

You can draw the balance sheet entirely from the trial balance. Note that the order of the general ledger is identical to that used in the balance sheet. The "net worth" section is the combination of the ending balance of net worth, plus current profits.

ABC Construction Company
Balance Sheet
December 31, 1988

Current Assets:		
Cash in Bank	$ 4,283	
Accounts Receivable	84,416	
Inventory	17,400	
Total Current Assets		$106,099
Long Term Assets:		
Trucks	$ 37,400	
Equipment	18,945	
Furniture	10,462	
Total	$ 66,807	
Less: Accumulated Depreciation	(36,300)	
Net Long-term Assets		$ 30,507
Total Assets		$136,606
Current Liabilities:		
Accounts Payable	$42,811	
Payroll Taxes Payable	2,701	
Notes Payable	4,175	
Total Current Liabilities		$ 49,687
Long-term Liabilities:		
Notes Payable		16,550
Total Liabilities		$ 66,237
Net Worth:		$ 70,369
Total Liabilities and Net Worth		$136,606

ABC Construction Company
Income Statement
For the year ending December 31, 1988

Sales		$473,010
Cost of Goods Sold:		
Inventory, 1-1-88	$ 19,100	
Materials Purchased	124,840	
Direct Labor	138,060	
Other Direct Costs	18,908	
Total	$300,908	
Less: Inventory, 12-31-88	17,400	
Cost of Goods Sold		$283,508
Gross Profit		$189,502
Variable Expenses:	$ 28,410	
Fixed Expenses	153,907	
Total Expenses		$182,317
Net Operating Profit		$ 7,185
Federal Income Tax		$ 1,078
Net Profit		$ 6,107

Question 2.

The income statement is also drawn from the-trial balance. The amount of profit must equal the derived total on the trial balance, and is also added to net worth on the balance sheet. The entry, "change inventory," represents the difference between opening and closing inventory levels.

ABC Construction Company
Cash Flow Statement
For the year ending December 31, 1988

Sources of Funds:			
Net Profit		$ 6,107	
Plus: Non-Cash Expenses		8,900	
Total		$ 15,007	
Sale of Long-Term Assets		-0-	
Total Sources of Funds			$ 15,007
Application of Funds:			
Purchase of Long-Term Assets		$ -0-	
Decrease in Notes Payable		10,481	
Total Applications of Funds			$ 10,481
Net Change			$ 4,526
Changes in Working Capital:			
	Last Year	This Year	Net Change
Cash	$13,902	$ 4,283	$ – 9,619
Accounts Receivable	79,632	84,416	4,784
Inventory	19,100	17,400	– 1,700
Accounts Payable	– 54,380	– 42,811	11,569
Payroll Taxes Payable	– 2,193	– 2,701	– 508
Notes Payable	– 4,175	– 4,175	-0-
Net	$51,886	$56,412	$ 4,526

Question 3.

You construct the cash flow statement from the differences in account balances. The top section of the statement (sources and applications) is prepared from changes in balance sheet accounts (including net worth, where the total year's profit is included). The bottom part, changes in working capital, consists of changes in current asset and current liability accounts. The top part's total must equal the bottom part's total.

Chapter 9

Question 1. Answer: d

Explanation: All three records must contain the breakdown of every check. If you have a large number of employees, you'll save a lot of time by using a write-once system, or an outside service, for payroll accounting.

Question 2. Answer: b

Explanation: The liability is the combination of those amounts withheld from the employees' checks, and the amount owed by the employer. You have to control and account for both for the account to remain in balance.

Question 3. Answer: a

Explanation: Payroll records are books of original entry. They're also subject to audit by federal and state taxing agencies and by your workers' compensation insurer.

Question 4. Answer: d

Explanation: Base your decision to establish a specialized account on whether you're having problems managing payroll through the general account. If you use an outside service, you may have to establish a separate account just to conform to the requirements of the service provider.

Question 5. Answer: c

Explanation: The payroll tax liability includes both withholding and employer's contributions.

Question 6.

See the completed worksheet at the bottom of the page:

Control note: This worksheet also serves to reconcile the current payroll. If you've filled out the worksheet correctly, the total of the under and over columns in each set will always equal the current gross payroll. In this case each of the three sets totals $11,200.

Chapter 10

Question 1. Answer: c

Explanation: While the term "depreciation" has several different meanings, the bookkeeping term refers strictly to the process of deducting a capital asset over several years.

Employees	SUI - $7,000		SDI - $21,900		FICA - $45,000	
	under	over	under	over	under	over
A	1,000		1,000		1,000	
B	500	700	1,200		1,200	
C		2,500	2,400	100	2,500	
D		2,500		2,500	2,500	
E		4,000		4,000	3,000	1,000
Totals	1,500	9,700	4,600	6,600	10,200	1,000

Question 2. Answer: d

Question 3. Answer: a

Explanation: Declining balance depreciation lets you deduct more in the early years. The benefit then declines. The ACRS prescribed method in all but the real estate classes is a combination of declining balance and straight-line depreciation.

Question 4. Answer: c

Explanation: You increase the expense account (to reduce profits) and decrease (credit) accumulated depreciation, which reduces the book value of capital assets.

Question 5. Answer: d

Explanation: No declining balance depreciation is allowed on commercial real estate.

Question 6.

Figure 10-2 gave percentages for each year under ACRS prescribed method depreciation. In the 5-year class, those percentages are:

Year	Percent
1	20.00
2	32.00
3	19.20
4	11.52
5	11.52
6	5.76

It takes six years to claim all the depreciation, since you're allowed only half the first year's rate in the first year. Apply these percentages to the $5,000 asset to develop depreciation under the prescribed method. Straight-line calculation consists of claiming one-fifth of the total for each of the five years, again limited to the allowed one-half deduction during the first year. See the worksheet at the bottom of the page.

Chapter 11

Question 1. Answer: c

Explanation: You separate a group of transactions into a subsidiary ledger to maintain better control. Your primary books — the general journal, receipts ledger, and general ledger — should be free of the kind of detail you need to track accounts receivable. The subsidiary books let you keep separate customer accounts, prepare monthly billings and track past-due accounts.

Question 2. Answer: d

Explanation: All three of the listed choices are correct. The isolation of details is a critical and valuable feature of subsidiary bookkeeping.

Question 3. Answer: a

Explanation: Each account requires its own card. Even if you only post one transaction during the month, or some customers only charge occasionally, combined records are inefficient and impractical.

Year	Prescribed Method	Straight-line Depreciation
1	$1,000	$ 500
2	1,600	1,000
3	960	1,000
4	576	1,000
5	576	1,000
6	288	500

Question 4. Answer: c

Explanation: Post as frequently as works best for you. If you have a relatively low volume, you might find it easiest to post both the ledger and the customer card at the same time. Do this daily, weekly, or twice per month, as required. Consider how frequently you and others need to review the customer accounts. Make your schedule suit your particular requirements.

Question 5. Answer: b

Explanation: Balance as often as necessary, depending on your transaction volume. This way you'll avoid the pressure of looking for errors when it's time to mail statements. And before you send out any statements, be sure that you balance completely so you know your bills are correct.

Question 6. Answer: d

Identify the one best source for consistent posting. That means the most dependable document that gives you all of the information you need. In most cases this will be an invoice. Decide which source document you'll use, then use it consistently.

Chapter 12

Question 1. Answer: c

Explanation: This system lets you see in advance what is due on each date.

Question 2. Answer: d

Explanation: All answers are correct. The owner can easily check the file at any time, which is an important benefit. (Owners often feel isolated because they don't understand how the books are set up.) If you have only a few transactions per month, the simplest procedure is always best. As a basic approach to recordkeeping, avoid setting up a permanent record if it's not needed.

Question 3. Answer: c

Explanation: Your periodic financial statements will reflect the true condition of your business under the accrual system. The key is to make the accrual and reversal procedure as simple and as manageable as possible.

Question 4. Answer: c

Explanation: The disadvantage of accrual accounting is the danger that not all the accruals will be reversed. An unreversed liability may remain on the books, and some payments can be entered twice. You can solve this problem by making a complete reversing entry at the beginning of the new month, then coding payments as you make them.

Chapter 13

Question 1. Answer: c

Explanation: By setting up a petty cash fund, you can both reconcile and verify cash expenses, summarizing cash activity in a way that's consistent with the rest of your bookkeeping system.

Question 2. Answer: d

Explanation: All answers are correct. Cash payments are legitimate business expenses, but they must be supported by a receipt. A petty cash fund is a convenience that also provides needed documentation.

Question 3. Answer: a

Explanation: In the general ledger, a petty cash fund is set up by writing a check from the general account. Then reimbursements are coded as debits to various accounts based on the supporting receipts, and a credit to the general cash account.

Question 4. Answer: d

Explanation: All answers are correct. The total amount in petty cash doesn't change. But there may be different levels of cash, postage and receipts. When you reimburse the fund, you replace receipts with cash.

Question 5. Answer: b

Explanation: The fund, which consists of cash, postage and receipts, should always equal the balance as established.

Question 6. Answer: c

Explanation: You reconcile petty cash whenever you replace receipts with cash. When the total of the fund is added up, it should equal the fund's established balance. The reconciliation serves a dual purpose. It not only verifies the total, but also breaks down the reimbursement by general ledger code.

Chapter 14

Question 1. Answer: c

Explanation: The purpose of job costing is to show profits or losses, and to control scheduling. In a sense, it's a budget for each job. The breakdown must be accurate, complete, and up to date to be of any use.

Question 2. Answer: b

Explanation: You can't calculate labor based only on the hourly rate paid to employees. Payroll taxes and payments made by the employer for union benefits add a significant factor to the total payroll cost.

Question 3. Answer: d

Explanation: Labor costs can be calculated several ways. The most efficient method will depend on your number of employees and jobs, and on whether or not an outside service provides you with a breakdown.

Question 4. Answer: b

Explanation: If you don't maintain an inventory, you can simply code each requisition to the appropriate job. But when you also have your own inventory, there are two possible sources for materials cost information. You need to account for direct purchases and subtractions from inventory. The procedure you use must allow for this.

Question 5. Answer: d

Explanation: You have several sources for direct costs. Your procedure should capture information from all sources, on an as-paid basis, and whenever costs are committed. The payments journal is the source for materials, subcontractor payments, licenses, freight, bonding, and other direct costs. The payroll account provides a breakdown of direct labor. And you'll use worksheets to allocate costs.

Question 6. Answer: d

Explanation: All answers apply.

Chapter 15

Question 1. Answer: c

Explanation: There are usually timing differences at some level, because deposits may be made after the statement cut-off date, or your payee's bank hasn't submitted checks for payment.

Question 2. Answer: d

Explanation: All answers are correct. You need a daily balance to monitor cash flow, time future payments and anticipate available cash. When you catch errors as you go, your month-end reconciliation will be easier.

Question 3. Answer: a

Explanation: The reconciliation is the way you identify the various adjustments required. It's complete when those adjustments have been made or identified.

Question 4. Answer: d

Explanation: Any of these adjustments may be necessary. You have to know exactly why an adjustment is necessary, and whether it's a bank problem or a checkbook problem.

Question 5. Answer: c

Explanation: All the answers are part of the reconciliation process. But to *complete* the process, you have to correct any errors in your books, and enter the adjustments.

Question 6. Answer: d

Explanation: All answers are correct. A journal entry is necessary to adjust the total disbursements for the previous month. And, the check is no longer outstanding.

Chapter 16

Question 1. Answer: d

Explanation: All answers are correct. The process of reconciling accounts helps control the quality of the general ledger, catch coding errors, identify the need for reversing entries, and be sure the ledger is accurate.

Question 2. Answer: b

Explanation: You do account reconciliations by comparing general ledger balances to details in subsidiary accounts, or by preparing analysis worksheets. The latter category includes the monthly bank account reconciliation.

Question 3. Answer: d

Explanation: The purpose of the account is to record transactions that can't be classified when they're recorded, or that you expect to reverse later (such as deposits.) Because each entry requires a separate reversal or recoding journal, it's quite likely that balances will accumulate in the suspense account unless you do a monthly reconciliation.

Question 4. Answer: d

Explanation: All choices are correct. One of the most important features of reconciliation is that it points out weaknesses or control problems in your bookkeeping procedure.

Question 5. Answer: a

Explanation: The purpose of summaries and worksheets is to define problems in accounts, or to verify the components and ending balances.

Question 6. Answer: a

Explanation: This file is a reminder for all recurring entries and for those journals you must prepare to reverse a previous posting. It is essential to identifying parts of the closing procedure that might otherwise be overlooked.

Chapter 17

Question 1. Answer: d

Explanation: All answers are correct. It saves you posting time. It works better in an automated system that manages numbers more efficiently than text. And if properly designed, it lets you break out sub-account and job cost information.

Question 2. Answer: c

Explanation: You find posting errors the same way, whether you use a code or not. You have to review the journals and posted transactions. With a code, you have the added benefit of identifying the type of transaction by its first digit.

Question 3. Answer: a

Explanation: The first digit of the account tells whether it's a balance sheet account (asset, liability, or net worth) or a profit and loss account (sales, direct costs, expenses). This feature will help you spot obvious coding errors.

Question 4. Answer: a

Explanation: A current asset is distinguished from a long-term asset. A variable expense is quickly distinguished from a fixed overhead account.

Question 5. Answer: d

Explanation: None of the answers apply. The purpose of the sub-account is to give you more detailed account information without the need for a complete subsidiary record. Thus, the use of sub-accounts is extremely limited. It won't be efficient in most cases for accounts receivable, inventory, or account analysis. It does apply to payroll tax liabilities, notes payable (when more than one loan is outstanding), and breakdowns of certain expense accounts.

Question 6. Answer: b

Explanation: Job cost information applies only to direct costs that can be identified by individual job. Attempting to assign overhead on this basis is arbitrary and provides no useful information; it only increases your workload.

Chapter 18

Question 1. Answer: d

Explanation: All answers apply. The books show every transaction and refer the summarized general ledger entry to the appropriate journal, and from there to the source document. The system also verifies the correct balance and posting of each account, as a feature of the double-entry method. And the information in the books is used to prepare accurate and up to date financial statements.

Question 2. Answer: c

Explanation: A bookkeeping system is of no value unless it's built on control. If it's only a collection of journals and ledgers, it doesn't do the most important job — staying in control of the money. Control must be the primary feature of the bookkeeping system.

Question 3. Answer: d

Explanation: All answers are correct. Embezzlement usually occurs when one person is allowed to handle cash and records without any checking by someone else, when the owner fails to supervise the books, or when the owner isn't aware of the need for control.

Question 4. Answer: a

Explanation: The bookkeeper's job should be very clearly defined and limited. It is to provide the owner with dependable, timely, controlled information. The other aspects of managing financial matters, such as planning, trend analysis, and projecting the future, are the responsibilities of the owner and his accountant.

Question 5. Answer: b

Explanation: The point where the bookkeeper's job ends and the accountant's begins is a clear one. If a function involves documentation, control, or proof of a transaction, it's a bookkeeping role. And if any analysis or interpretation is involved, that's where the accountant's job begins.

Question 6. Answer: c

Explanation: Before you buy a computer system, you must do some research and price comparison. But the most important step is to evaluate your needs. This is a step that's often overlooked. For automation to be of benefit, it has to save you money and make your job easier.

GLOSSARY

A

Account analysis The investigation of balances and transactions in a general ledger to determine the contents of an account and to verify account codes and posting accuracy.

Accountant A professional who analyzes financial information, advises management, and interprets trends for a business owner. These services are based on the information in the books and records, both historical and current and on projections of future activity.

Accounts payable Business costs and expenses that are due and payable within the current period.

Accounts receivable Money due from customers who have purchased goods or services on credit.

Accrual basis A method of reporting transactions based on the accrual of income or incurring of costs and expenses.

Adjusting journals Journal entries to correct errors or reverse estimates.

Amortization The gradual reduction of an account over time, for instance, organizational expenses which are written off over 60 months.

Asset A property owned by a business. Assets may be current (cash or convertible to cash within one year); long-term (capital assets, such as equipment and machinery); prepaid (such as the value of unamortized organizational expenses or insurance premiums); or intangible (such as goodwill).

Audit trail The documentation of transactions that allows an auditor or accountant to verify the accuracy of information in the books. The trail follows a source document through the system of journal entries, then to the general ledger.

B

Bad debt Business expense resulting when an account receivable cannot be collected.

Balance sheet A financial statement that reports the status, as of a specific date, of assets, liabilities, and net worth.

Bank statement A monthly report from the bank that summarizes all debits (checks and bank charges) and credits (deposits) recorded in a checking account.

Bookkeeper A person who posts and balances all financial records for a business. The bookkeeper is also responsible for the audit trail, and provides financial information to the owner and the accountant.

Books of final entry The general ledger of a business.

Books of original entry One of several journals in the bookkeeping system. All information in the general ledger is posted from the receipts, disbursements, and general journals, or from subsidiary journals representing one of the three books of original entry.

C

Capital The invested value of ownership in a business. Capitalization may consist of both equity (capital investment by the owners) and debt (borrowed money).

Cash basis An accounting method in which transactions are posted to the books only when money changes hands.

Cash flow statement A financial statement that shows for a period of time (usually one year), how the business acquired and applied cash. This includes a breakdown of long-term assets and liabilities, net profits on a cash basis, and a summary of changes in working capital (current assets and liabilities).

Cash receipts journal A book of original entry in which cash receipts and sales are posted. May include accounts receivable information.

Chart of accounts A coded summary of all accounts in the general ledger. The purpose of coding is to reduce the writing involved in preparing and posting journals.

Closing the books The process of ending a month or a year, for the purpose of balancing and preparing financial statements. The monthly closing process involves a trial balance and worksheet-based adjusting entries. At the end of the year, all profit and loss accounts (sales, costs, and expenses) are reversed to zero balances, and the net profit or loss is transferred to the capital account.

Comparative statement A financial statement in which the current balances of accounts are shown in comparison to a previous period.

Consolidated statement A financial statement in which the results of subsidiaries, divisions, or profit and loss from separate lines of business are combined into a single report.

Corporation A company owned jointly by stockholders. The combined shares outstanding represent the total ownership, and the corporation operates as a single entity. Ownership may be transferred by the sale of stock, without disturbing the continuing operation of the business.

Cost accounting The breakdown of costs and, in some cases, expenses, by cost centers such as jobs. The purpose is to evaluate profitability of the cost centers on an individual basis.

Cost of goods sold The amount of money spent for materials, labor, and other costs directly related to specific jobs.

Credit (a) A right-sided entry in the double-entry bookkeeping system. (b) A method of selling in which the customer is allowed to pay for merchandise at a date later than the actual purchase.

Current assets Property of a business that is either cash or is convertible to cash within one year. Current assets includes cash, accounts receivable, and inventory.

Current liabilities All debts of a business that will become due and payable within one year, including the next twelve months' payments on notes payable.

Current ratio A comparison between current assets and current liabilities. The ratio is expressed by the fraction that results when you divide the former by the latter. A ratio of 2 to 1 or better is considered a sign that working capital is adequate.

D

Debit A left-sided entry in the double-entry bookkeeping system.

Deferred expenses Money spent before the expense has been incurred. The transaction is entered to an asset account and reversed when the actual expense is recorded.

Deferred income A credit-balance account included in the liability section of the balance sheet. Holds deposit receipts from customers. The deferred credits are reversed when income is actually earned and the customer is billed.

Depreciation The gradual writing off of the cost of long-term assets, over a specific recovery period. An entry is made each year to charge the depreciation expense account and reduce the value of the asset.

Direct labor The cost of payroll for employees working on specific jobs. In contrast, salaries and wages of office employees and owners aren't directly assigned to jobs.

Disbursements journal A permanent record and book of original entry. The journal lists all payments by ledger account. Monthly totals are a credit to cash and debits to various other accounts.

Double-entry books Permanent record of business transactions in which every entry contains both a debit and a credit. These are not positive and negative entries, but left- and right-sided ones.

E

Earned income Income earned within the current month. Income may be received in advance of the earned period (deferred income); in the same period earned; or after the earned period.

Expenses Money spent for variable or fixed overhead, as distinguished from direct costs.

F

Fixed assets Capital assets of the business that are subject to depreciation (machinery, equipment, autos and trucks, for example) or that are not depreciable (land, for example). The value of long-term assets is reduced by annual depreciation. Same as long-term assets.

Fixed expenses Expenses that remain unchanged regardless of sales volume, such as rent

G

General journal A journal used for recording adjustments, non-cash transactions, and other entries not appropriate to the receipts or disbursements journals.

General ledger The book of final entry in the bookkeeping system, where all transactions are summarized for each month, and ending balances are used to prepare financial statements.

Goodwill An intangible asset representing the perceived value of a company, based on reputation and name association among customers and the community.

Gross margin The percentage that gross profit represents of gross sales. It is computed by dividing gross profit by gross sales.

Gross profit Profit after direct costs are subtracted from sales, but before deduction of expenses.

I

Imprest petty cash A method of keeping records for cash expenses. A fund is set up with a specific amount, and, as it's depleted by reimbursements for expenses paid, the fund is reimbursed from the general cash account.

Income statement A financial statement, prepared for a specific period of time (usually one year) that summarized the period's sales, cost of goods sold, expenses, and net profit or loss.

Inventory Materials stored in a warehouse or yard, to be used on future jobs.

J

Job cost records Records of direct costs by job. The purpose is to (a) evaluate profit and loss by job; (b) spot emerging trends before problems develop; and (c) track the scheduled forecast of the job itself, and prepare progress billings.

L

Liability A debt of a company, due within one year (current) or beyond the next 12 months (long-term).

Long-term assets Capital assets of the business that are subject to depreciation (machinery, equipment, autos and trucks, for example) or that are not depreciable (land, for example). The value of long-term assets is reduced by annual depreciation. Also called *Fixed assets.*

Long-term liabilities Debts of the business that are due and payable beyond the coming 12 months.

M

Materials A direct cost for materials purchased for a specific job; or the value of materials removed from inventory and used on that job.

N

Net income The amount of money remaining after you subtract all costs and expenses from total sales. Net income may be expressed either on a before-tax or after-tax basis.

Notes payable Money due to a lender.

Notes receivable An asset for the amount due from another person or company, secured by a promissory note.

O

Organizational expenses An asset that is amortized over a period of time, rather than written off during the year money was spent. Organizational expenses are those spent before the company opened its doors for business.

Overhead expenses Expenses that don't vary with sales volume.

P

Partnership A proprietary business (not a corporation) with two or more owners. Each is self-employed and is fully liable for the debts of the business, and each is responsible for prepaying income and social security taxes.

Payroll account A special checking account for paying employee wages and related taxes.

Petty cash fund A fund used for expenses paid with cash, not checks.

Prepaid assets The value of expenses paid in one year but applicable to two or more years, such as a 36 month insurance premium.

Profit and loss account Part of the net worth section on the balance sheet, where the annual profit or loss is posted. At the end of the year, all sales, cost, and expense accounts are reversed to a zero balance, and the net amount is posted to profit and loss. A profit increases net worth, and a loss reduces it.

Purchases journal A record of accrued material purchases, used with a requisitioning and inventory control system.

R

Receipts journal A record of sales, and in some instances, payments on account.

Reconciliation The proof of an account's balance, including identification and correction of errors.

Reserve for bad debts A credit-balance account in the current asset section of the balance sheet. This reserve is established to allow for future bad debts. Increases in the reserve reduce current assets.

Reserve for depreciation Also called accumulated depreciation, this is a reduction of the fixed asset section of the balance sheet. It's the credit that accompanies depreciation expense entries.

Retained earnings The accumulated balance of net profits kept in the business over a period of years, less absorbed net losses.

S

Single-entry books Books that contain one entry for each transaction. In most applications of this system, the record is one of cash received or spent; the entry itself identifies income, costs or expenses.

Sole proprietorship A business with one owner. Profits may be withdrawn or left to accumulate in the business, and profits are taxed whether they are drawn or not.

Source document An invoice, receipt, statement, voucher, or other document that supports income or the payment of a business cost or expense.

Subsidiary ledger A section of the books used to manage a particular grouping of transactions, such as accounts receivable.

Supplementary schedule A detailed list attached to a financial statement that explains a summary entry on the statement itself.

T

Trial balance A worksheet summarizing and proving the balances of all accounts in the general ledger. A preliminary step to the preparation of financial statements.

V

Variable expenses Expenses that may rise or fall in line with changes in sales volume, such as auto and truck expenses, but not to the same degree as direct costs.

W

Working capital Current assets minus current liabilities. This is an indication of the relative health of business cash flow.

INDEX

Other Practical References

National Construction Estimator
Current building costs in dollars and cents for residential, commercial and industrial construction. Prices for every commonly used building material, and the proper labor cost associated with installation of the material. Everything figured out to give you the "in place" cost in seconds. Many time-saving rules of thumb, waste and coverage factors and estimating tables are included. **544 pages, 8½ x 11, $19.50. Revised annually.**

Building Cost Manual
Square foot costs for residential, commercial, industrial, and farm buildings. In a few minutes you work up a reliable budget estimate based on the actual materials and design features, area, shape, wall height, number of floors and support requirements. Most important, you include all the important variables that can make any building unique from a cost standpoint. **240 pages, 8½ x 11, $14.00. Revised annually**

Berger Building Cost File
Labor and material costs needed to estimate major projects: shopping centers and stores, hospitals, educational facilities, office complexes, industrial and institutional buildings, and housing projects. All cost estimates show both the manhours required and the typical crew needed so you can figure the price and schedule the work quickly and easily. **304 pages, 8½ x 11, $30.00. Revised annually**

Estimating Home Building Costs
Estimate every phase of residential construction from site costs to the profit margin you should include in your bid. Shows how to keep track of manhours and make accurate labor cost estimates for footings, foundations, framing and sheathing finishes, electrical, plumbing and more. Explains the work being estimated and provides sample cost estimate worksheets with complete instructions for each job phase. **320 pages, 5½ x 8½, $17.00**

Construction Estimating Reference Data
Collected in this single volume are the building estimator's 300 most useful estimating reference tables. Labor requirements for nearly every type of construction are included: site work, concrete work, masonry, steel, carpentry, thermal & moisture protection, doors and windows, finishes, mechanical and electrical. Each section explains in detail the work being estimated and gives the appropriate crew size and equipment needed. **368 pages, 11 x 8½, $20.00**

Estimating Tables for Home Building
Produce accurate estimates in minutes for nearly any home or multi-family dwelling. This handy manual has the tables you need to find the quantity of materials and labor for most residential construction. Includes overhead and profit, how to develop unit costs for labor and materials and how to be sure you've considered every cost in the job. **336 pages, 8½ x 11, $21.50**

Electrical Construction Estimator
If you estimate electrical jobs, this is your guide to current material costs, reliable manhour estimates per unit, and the total installed cost for all common electrical work: conduit, wire, boxes, fixtures, switches, outlets, loadcenters, panelboards, raceway, duct, signal systems, and more. Explains what every estimator should know before estimating each part of an electrical system. **416 pages, 8½ x 11, $25.00. Revised annually**

Contractor's Year-Round Tax Guide
How to set up and run your construction business to minimize taxes: corporate tax strategy and how to use it to your advantage, and what you should be aware of in contracts with others. Covers tax shelters for builders, write-offs and investments that will reduce your taxes, accounting methods that are best for contractors, and what the I.R.S. allows and what it often questions. **192 pages, 8½ x 11, $16.50**

Builder's Guide to Construction Financing
Explains how and where to borrow the money to buy land and build homes and apartments: conventional loan sources, loan brokers, private lenders, purchase money loans, and federally insured loans. How to shop for financing, get the valuation you need, comply with lending requirements, and handle liens. **304 pages, 5½ x 8½, $15.25**

Computers: The Builder's New Tool
Shows how to avoid costly mistakes and find the right computer system for your needs. Takes you step-by-step through each important decision, from selecting the software to getting your equipment set up and operating. Filled with examples, checklists and illustrations, including case histories describing experiences other contractors have had. If you're thinking about putting a computer in your construction office, you should read this book before buying anything. **192 pages, 8½ x 11, $17.75**

Builder's Guide to Accounting Revised
Step-by-step, easy to follow guidelines for setting up and maintaining an efficient record keeping system for your building business. Not a book of theory, this practical, newly-revised guide to all accounting methods shows how to meet state and federal accounting requirements, including new depreciation rules, and explains what the tax reform act of 1986 can mean to your business. Full of charts, diagrams, blank forms, simple directions and examples. **304 pages, 8½ x 11, $17.25**

Cost Records for Construction Estimating
How to organize and use cost information from jobs just completed to make more accurate estimates in the future. Explains how to keep the cost records you need to reflect the time spent on each part of the job. Shows the best way to track costs for sitework, footing, foundations, framing, interior finish, siding and trim, masonry, and subcontract expense. Provides sample forms. **208 pages, 8½ x 11, $15.75**

Contractor's Survival Manual
How to survive hard times in construction and take full advantage of the profitable cycles. Shows what to do when the bills can't be paid, finding money and buying time, transferring debt, and all the alternatives to bankruptcy. Explains how to build profits, avoid problems in zoning and permits, taxes, time-keeping, and payroll. Unconventional advice includes how to invest in inflation, get high appraisals, trade and postpone income, and how to stay hip-deep in profitable work. **160 pages, 8½ x 11, $16.75**

Contractor's Guide to the Building Code
Explains in plain English exactly what the Uniform Building Code requires and shows how to design and construct residential and light commercial buildings that will pass inspection the first time. Suggests how to work with the inspector to minimize construction costs, what common building short cuts are likely to be cited, and where exceptions are granted. **312 pages, 5½ x 8½, $16.25**

Builder's Office Manual, Revised
Explains how to create routine ways of doing all the things that must be done in every construction office — in the minimum time, at the lowest cost, and with the least supervision possible: Organizing the office space, establishing effective procedures and forms, setting priorities and goals, finding and keeping an effective staff, getting the most from your record-keeping system (whether manual or computerized). Loaded with practical tips, charts and sample forms for your use. **192 pages, 8½ x 11, $15.50**

Audio Tape: Construction Field Supervision
No project is better than the supervisor running it. That's why hundreds of foremen, supervisors, and contractors have used Carl Bauman's two-day workshops to sharpen field supervision skills. Now this popular seminar is available on audiotape at a fraction of the cost of attending in person. If poor job supervision is costing you money, order this package. **Six 90-minute tapes and 178 page workbook, only $95.00**

Blueprint Reading for the Building Trades
How to read and understand construction documents, blueprints, and schedules. Includes layouts of structural, mechanical and electrical drawings, how to interpret sectional views, how to follow diagrams; plumbing, HVAC and schematics, and common problems experienced in interpreting construction specifications. This book is your course for understanding and following construction documents. **192 pages, 5½ x 8½, $11.25**

HVAC Contracting
Your guide to setting up and running a successful HVAC contracting company. Shows how to plan and design all types of systems for maximum efficiency and lowest cost — and explains how to sell your customers on the designs you propose. Describes the right way to use all the instruments, equipment and reference materials essential to HVAC contracting. Includes a full chapter on estimating, bidding, and contract procedure. **256 pages, 8½ x 11, $24.50**

Residential Wiring
Shows how to install rough and finish wiring in both new construction and alterations and additions. Complete instructions are included on troubleshooting and repairs. Every subject is referenced to the 1987 National Electrical Code, and over 24 pages of the most needed NEC tables are included to help you avoid errors so your wiring passes inspection — the first time. **352 pages, 5½ x 8½, $18.25**

Contractor's Growth and Profit Guide
Step-by-step instructions for planning growth and prosperity in a construction contracting or subcontracting company. Explains how to prepare a business plan: selecting reasonable goals, drafting a market expansion plan, making income forecasts and expense budgets, and projecting cash flow. Here you will learn everything required by most lenders and investors, as well as solid knowledge for better organizing your business. **336 pages, 5½ x 8½, $19.00**

Carpentry for Residential Construction
How to do professional quality carpentry work in homes and apartments. Illustrated instructions show you everything from setting batter boards to framing floors and walls, installing floor, wall and roof sheathing, and applying roofing. Covers finish carpentry, also: How to install each type of cornice, frieze, lookout, ledger, fascia and soffit; how to hang windows and doors; how to install siding, drywall and trim. Each job description includes the tools and materials needed, the estimated manhours required, and a step-by-step guide to each part of the task. **400 pages, 5½ x 8½, $19.75**

Carpentry in Commercial Construction
Covers forming, framing, exteriors, interior finish and cabinet installation in commercial buildings: designing and building concrete forms, selecting lumber dimensions, grades and species for the design load, what you should know when installing materials selected for their fire rating or sound transmission characteristics, and how to plan and organize the job to improve production. Loaded with illustrations, tables, charts and diagrams. **272 pages, 5½ x 8½, $19.00**

Excavation & Grading Handbook, Revised
Explains how to handle all excavation, grading, compaction, paving and pipeline work: setting cut and fill stakes (with both bubble and laser levels), working in rock, unsuitable material or mud, passing compaction tests, trenching around utility lines, setting grade pins and string line, removing or laying asphaltic concrete, widening roads, cutting channels, installing water, sewer and drainage pipe. This is the completely revised edition of the popular guide used by over 25,000 excavation contractors and supervisors. **384 pages, 5½ x 8½, $22.75**

Construction Surveying and Layout
A practical guide to simplified construction surveying: How land is divided, how to use a transit and tape to find a known point, how to draw an accurate survey map from your field notes, how to use topographic surveys, and the right way to level and set grade. You'll learn how to make a survey for any residential or commercial lot, driveway, road, or bridge — including how to figure cuts and fills and calculate excavation quantities. If you've been wanting to make your own surveys, or just read and verify the accuracy of surveys made by others, you should have this guide. **256 pages, 5½ x 8½, $19.25**

Carpentry Layout
Explains the easy way to figure: Cuts for stair carriages, treads and risers. Lengths for common, hip and jack rafters. Spacing for joists, studs, rafters and pickets. Layout for rake and bearing walls. Shows how to set foundation corner stakes — even for a complex home on a hillside. Practical examples show how to use a hand-held calculator as a powerful layout tool. Written in simple language any carpenter can understand. **240 pages, 5½ x 8½, $16.25**

Pipe and Excavation Contracting
What you need to know to succeed as a pipe and excavation contractor. How to read plans and compute quantities for both trench and surface excavation, figure crew and equipment productivity rates, estimate unit costs, bid the work, and get the bonds you need. Explains what equipment will deliver maximum productivity for each job, how to lay all types of water and sewer pipe, and how to switch your business to excavation work when you don't have pipe contracts. Covers asphalt and rock removal, working on steep slopes or in high groundwater, and how to avoid the pitfalls that can wipe out your profits on any job. **400 pages, 5½ x 8½, $23.50**

Wood Frame House Construction
From the layout of the outer walls, excavation and formwork, to finish carpentry, and painting, every step of construction is covered in detail with clear illustrations and explanations. Everything the builder needs to know about framing, roofing, siding, insulation and vapor barrier, interior finishing, floor coverings, and stairs ... complete step by step "how to" information on what goes into building a frame house. **240 pages, 8½ x 11, $14.25. Revised edition.**

How to Sell Remodeling
Proven, effective sales methods for repair and remodeling contractors: finding qualified leads, making the sales call, identifying what your prospects really need, pricing the job, arranging financing, and closing the sale. Explains how to organize and staff a sales team, how to bring in the work to keep your crews busy and your business growing, and much more. Includes blank forms, tables, and charts. **240 pages, 8½ x 11, $17.50**

Handbook of Construction Contracting Vol. 1 & 2
Volume 1: Everything you need to know to start and run your construction business; the pros and cons of each type of contracting, the records you'll need to keep, and how to read and understand house plans and specs to find any problems before the actual work begins. All aspects of construction are covered in detail, including all-weather wood foundations, practical math for the job-site, and elementary surveying. **416 pages, 8½ x 11, $21.75**

Volume 2: Everything you need to know to keep your construction business profitable; different methods of estimating, keeping and controlling costs, estimating excavation, concrete, masonry, rough carpentry, roof covering, insulation, doors and windows, exterior finish, specialty finishes, scheduling work flow, managing workers, advertising and sales, spec building and land development and selecting the best legal structure for your business. **320 pages, 8½ x 11, $24.75**

Carpentry Estimating
Simple, clear instructions show you how to take off quantities and figure costs for all rough and finish carpentry. Shows how much overhead and profit to include, how to convert piece prices to MBF prices or linear foot prices, and how to use the tables included to quickly estimate manhours. All carpentry is covered: floor joists, exterior and interior walls and finishes, ceiling joists and rafters, stairs, trim, windows, doors, and much more. Includes sample forms, checklists, and the author's factor worksheets to save you time and help prevent errors. **320 pages, 8½ x 11, $25.50**

Drywall Contracting
How to do professional quality drywall work, how to plan and estimate each job, and how to start and keep your drywall business thriving. Covers the eight essential steps in making any drywall estimate, how to achieve the six most commonly-used surface treatments, how to work with metal studs, and how to solve and prevent most common drywall problems. **288 pages, 5½ x 8½, $18.25**

Construction Superintending
Explains what the "super" should do during every job phase from taking bids to project completion on both heavy and light construction: excavation, foundations, pilings, steelwork, concrete and masonry, carpentry, plumbing, and electrical. Explains scheduling, preparing estimates, record keeping, dealing with subcontractors, and change orders. Includes the charts, forms, and established guidelines every superintendent needs. **240 pages, 8½ x 11, $22.00**

Paint Contractor's Manual
How to start and run a profitable paint contracting company: getting set up and organized to handle volume work, avoiding the mistakes most painters make, getting top production from your crews and the most value from your advertising dollar. Shows how to estimate all prep and painting. Loaded with manhour estimates, sample forms, contracts, charts, tables and examples you can use. **224 pages, 8½ x 11, $19.25**

Craftsman Book Company
6058 Corte del Cedro, P.O. Box 6500
Carlsbad, CA 92008 - 0992
FAX No. (619) 438-0398

In a hurry?
We accept phone orders charged to your MasterCard, Visa or American Express
Call (619) 438-7828

Mail Orders
We pay shipping when your check covers your order in full.

Name ____________

Company ____________

Address ____________

City ____________ **State** ____________ **Zip** ____________

Send check or money order
Total Enclosed ____________ **(In California add 6% tax)**
If you prefer, use your:
☐ **Visa,** ☐ **MasterCard or** ☐ **American Express**

Card # ____________

Expiration date ____________ **Initials** ______

10 Day Money Back GUARANTEE

- ☐ 95.00 Audio: Construction Field Supervision
- ☐ 30.00 Berger Building Cost File
- ☐ 11.25 Blueprint Reading for Building Trades
- ☐ 17.25 Builder's Guide to Accounting Revised
- ☐ 15.25 Builder's Guide to Const. Financing
- ☐ 15.50 Builder's Office Manual Revised
- ☐ 14.00 Building Cost Manual
- ☐ 25.50 Carpentry Estimating
- ☐ 19.75 Carpentry for Residential Construction
- ☐ 19.00 Carpentry in Commercial Construction
- ☐ 16.25 Carpentry Layout
- ☐ 17.75 Computers: The Builder's New Tool
- ☐ 20.00 Const. Estimating Ref. Data
- ☐ 22.00 Construction Superintending
- ☐ 19.25 Construction Surveying and Layout
- ☐ 19.00 Contractor's Growth and Profit Guide
- ☐ 16.25 Contractor's Guide to the Building Code
- ☐ 16.75 Contractor's Survival Manual
- ☐ 16.50 Contractor's Year-Round Tax Guide
- ☐ 15.75 Cost Records for Const. Estimating
- ☐ 18.25 Drywall Contracting
- ☐ 25.00 Electrical Construction Estimator
- ☐ 17.00 Estimating Home Building Costs
- ☐ 21.50 Estimating Tables for Home Building
- ☐ 22.75 Excavation & Grading Handbook, Revised
- ☐ 21.75 Handbook of Construction Contracting Vol. 1
- ☐ 24.75 Handbook of Construction Contracting Vol. 2
- ☐ 17.50 How to Sell Remodeling
- ☐ 24.50 HVAC Contracting
- ☐ 19.50 National Construction Estimator
- ☐ 19.25 Paint Contractor's Manual
- ☐ 23.50 Pipe & Excavation Contracting
- ☐ 18.25 Residential Wiring
- ☐ 14.25 Wood-Frame House Construction
- ☐ 19.75 Bookkeeping For Builders

These books are tax deductible when used to improve or maintain your professional skill.

10 DAY MONEY BACK GUARANTEE

bfb card

Craftsman Book Company
6058 Corte del Cedro
P. O. Box 6500
Carlsbad, CA 92008

Craftsman

In a hurry?
We accept phone orders charged to your MasterCard, Visa or Am. Ex.
Call (619) 438-7828

Name (Please print clearly) ____________

Company ____________

Address ____________

City/State/Zip ____________

Total Enclosed ____________ (In California add 6% tax)

Use your ☐ Visa ☐ MasterCard ☐ Am. Ex.

Card # ____________

Exp. date ____________ Initials ______

- ☐ 95.00 Audio: Construction Field Sup.
- ☐ 17.50 Basic Plumbing with Illust.
- ☐ 30.00 Berger Building Cost File
- ☐ 11.25 Blprt Read. for Blding Trades
- ☐ 19.75 Bookkeeping For Builders
- ☐ 17.25 Builder's Guide to Accting. Rev.
- ☐ 15.25 Blder's Guide to Const. Fin.
- ☐ 15.50 Builder's Office Manual Revised
- ☐ 14.00 Building Cost Manual
- ☐ 11.75 Building Layout
- ☐ 25.50 Carpentry Estimating
- ☐ 19.75 Carp. for Residential Const.
- ☐ 19.00 Carp. in Commercial Const.
- ☐ 16.25 Carpentry Layout
- ☐ 17.75 Computers: Blder's New Tool
- ☐ 14.50 Concrete and Formwork
- ☐ 20.50 Concrete Const. & Estimating
- ☐ 20.00 Const. Estimating Ref. Data
- ☐ 22.00 Construction Superintending
- ☐ 19.25 Const. Surveying & Layout
- ☐ 19.00 Cont. Growth & Profit Guide
- ☐ 16.25 Cont. Guide to the Blding Code
- ☐ 16.75 Contractor's Survival Manual
- ☐ 16.50 Cont.Year-Round Tax Guide
- ☐ 15.75 Cost Rec. for Const. Est.
- ☐ 9.50 Dial-A-Length Rafterrule
- ☐ 18.25 Drywall Contracting
- ☐ 13.75 Electrical Blueprint Reading
- ☐ 25.00 Electrical Const. Estimator
- ☐ 19.00 Estimating Electrical Const.
- ☐ 17.00 Estimating Home Blding Costs
- ☐ 17.25 Estimating Plumbing Costs
- ☐ 21.50 Esti. Tables for Home Building
- ☐ 22.75 Exca. & Grading Handbook, Rev.
- ☐ 9.25 E-Z Square
- ☐ 10.50 Finish Carpentry
- ☐ 21.75 Hdbk of Const. Cont. Vol. 1
- ☐ 24.75 Hdbk of Const. Cont. Vol. 2
- ☐ 16.50 Hdbk of Modern Elect. Wiring
- ☐ 15.00 Home Wiring: Imp., Ext., Repairs
- ☐ 17.50 How to Sell Remodeling
- ☐ 24.50 HVAC Contracting
- ☐ 20.25 Manual of Elect. Contracting
- ☐ 19.75 Manual of Prof. Remodeling
- ☐ 13.50 Masonry & Concrete Const.
- ☐ 26.50 Masonry Estimating
- ☐ 19.50 National Const. Estimator
- ☐ 23.75 Op. the Tractor-Loader-Backhoe
- ☐ 23.50 Pipe & Excavation Contracting
- ☐ 19.25 Paint Contractor's Manual
- ☐ 21.25 Painter's Handbook
- ☐ 13.00 Plan. and Design. Plumbing Sys.
- ☐ 21.00 Plumber's Exam Prep. Guide
- ☐ 18.00 Plumber's Handbook Revised
- ☐ 14.25 Rafter Length Manual
- ☐ 18.50 Remodeler's Handbook
- ☐ 11.50 Residential Electrical Design
- ☐ 16.75 Residential Electrician's Hdbk.
- ☐ 18.25 Residential Wiring
- ☐ 22.00 Roof Framing
- ☐ 14.00 Roofers Handbook
- ☐ 16.00 Rough Carpentry
- ☐ 21.00 Running Your Remodeling Bus.
- ☐ 24.00 Spec Builder's Guide
- ☐ 13.75 Stair Builder's Handbook
- ☐ 15.50 Video: Asphalt Shingle Roofing
- ☐ 15.50 Video: Bathroom Tile
- ☐ 24.75 Video: Drywall Contracting 1
- ☐ 24.75 Video: Drywall Contracting 2
- ☐ 15.50 Video: Electrical Wiring
- ☐ 15.50 Video: Exterior Painting
- ☐ 15.50 Video: Finish Carpentry
- ☐ 15.50 Video: Hanging An Exterior Door
- ☐ 15.50 Video: Int. Paint & Wallpaper
- ☐ 15.50 Video: Kitchen Renovation
- ☐ 24.75 Video: Paint Contractor's 1
- ☐ 24.75 Video: Paint Contractor's 2
- ☐ 15.50 Video: Plumbing
- ☐ 80.00 Video: Roof Framing 1
- ☐ 80.00 Video: Roof Framing 2
- ☐ 15.50 Video: Rough Carpentry
- ☐ 15.50 Video: Windows & Doors
- ☐ 15.50 Video: Wood Siding
- ☐ 7.50 Visual Stairule
- ☐ 14.25 Wood-Frame House Const.

bfb card

Craftsman Book Company
6058 Corte del Cedro
P. O. Box 6500
Carlsbad, CA 92008

Craftsman

In a hurry?
We accept phone orders charged to your MasterCard, Visa or Am. Ex.
Call (619) 438-7828

Name (Please print clearly) ____________

Company ____________

Address ____________

City/State/Zip ____________

Total Enclosed ____________ (In California add 6% tax)

Use your ☐ Visa ☐ MasterCard ☐ Am. Ex.

Card # ____________

Exp. date ____________ Initials ______

- ☐ 95.00 Audio: Construction Field Sup.
- ☐ 17.50 Basic Plumbing with Illust.
- ☐ 30.00 Berger Building Cost File
- ☐ 11.25 Blprt Read. for Blding Trades
- ☐ 19.75 Bookkeeping For Builders
- ☐ 17.25 Builder's Guide to Accting. Rev.
- ☐ 15.25 Blder's Guide to Const. Fin.
- ☐ 15.50 Builder's Office Manual Revised
- ☐ 14.00 Building Cost Manual
- ☐ 11.75 Building Layout
- ☐ 25.50 Carpentry Estimating
- ☐ 19.75 Carp. for Residential Const.
- ☐ 19.00 Carp. in Commercial Const.
- ☐ 16.25 Carpentry Layout
- ☐ 17.75 Computers: Blder's New Tool
- ☐ 14.50 Concrete and Formwork
- ☐ 20.50 Concrete Const. & Estimating
- ☐ 20.00 Const. Estimating Ref. Data
- ☐ 22.00 Construction Superintending
- ☐ 19.25 Const. Surveying & Layout
- ☐ 19.00 Cont. Growth & Profit Guide
- ☐ 16.25 Cont. Guide to the Blding Code
- ☐ 16.75 Contractor's Survival Manual
- ☐ 16.50 Cont.Year-Round Tax Guide
- ☐ 15.75 Cost Rec. for Const. Est.
- ☐ 9.50 Dial-A-Length Rafterrule
- ☐ 18.25 Drywall Contracting
- ☐ 13.75 Electrical Blueprint Reading
- ☐ 25.00 Electrical Const. Estimator
- ☐ 19.00 Estimating Electrical Const.
- ☐ 17.00 Estimating Home Blding Costs
- ☐ 17.25 Estimating Plumbing Costs
- ☐ 21.50 Esti. Tables for Home Building
- ☐ 22.75 Exca. & Grading Handbook, Rev.
- ☐ 9.25 E-Z Square
- ☐ 10.50 Finish Carpentry
- ☐ 21.75 Hdbk of Const. Cont. Vol. 1
- ☐ 24.75 Hdbk of Const. Cont. Vol. 2
- ☐ 16.50 Hdbk of Modern Elect. Wiring
- ☐ 15.00 Home Wiring: Imp., Ext., Repairs
- ☐ 17.50 How to Sell Remodeling
- ☐ 24.50 HVAC Contracting
- ☐ 20.25 Manual of Elect. Contracting
- ☐ 19.75 Manual of Prof. Remodeling
- ☐ 13.50 Masonry & Concrete Const.
- ☐ 26.50 Masonry Estimating
- ☐ 19.50 National Const. Estimator
- ☐ 23.75 Op. the Tractor-Loader-Backhoe
- ☐ 23.50 Pipe & Excavation Contracting
- ☐ 19.25 Paint Contractor's Manual
- ☐ 21.25 Painter's Handbook
- ☐ 13.00 Plan. and Design. Plumbing Sys.
- ☐ 21.00 Plumber's Exam Prep. Guide
- ☐ 18.00 Plumber's Handbook Revised
- ☐ 14.25 Rafter Length Manual
- ☐ 18.50 Remodeler's Handbook
- ☐ 11.50 Residential Electrical Design
- ☐ 16.75 Residential Electrician's Hdbk.
- ☐ 18.25 Residential Wiring
- ☐ 22.00 Roof Framing
- ☐ 14.00 Roofers Handbook
- ☐ 16.00 Rough Carpentry
- ☐ 21.00 Running Your Remodeling Bus.
- ☐ 24.00 Spec Builder's Guide
- ☐ 13.75 Stair Builder's Handbook
- ☐ 15.50 Video: Asphalt Shingle Roofing
- ☐ 15.50 Video: Bathroom Tile
- ☐ 24.75 Video: Drywall Contracting 1
- ☐ 24.75 Video: Drywall Contracting 2
- ☐ 15.50 Video: Electrical Wiring
- ☐ 15.50 Video: Exterior Painting
- ☐ 15.50 Video: Finish Carpentry
- ☐ 15.50 Video: Hanging An Exterior Door
- ☐ 15.50 Video: Int. Paint & Wallpaper
- ☐ 15.50 Video: Kitchen Renovation
- ☐ 24.75 Video: Paint Contractor's 1
- ☐ 24.75 Video: Paint Contractor's 2
- ☐ 15.50 Video: Plumbing
- ☐ 80.00 Video: Roof Framing 1
- ☐ 80.00 Video: Roof Framing 2
- ☐ 15.50 Video: Rough Carpentry
- ☐ 15.50 Video: Windows & Doors
- ☐ 15.50 Video: Wood Siding
- ☐ 7.50 Visual Stairule
- ☐ 14.25 Wood-Frame House Const.

Craftsman Book Company, 6058 Corte Del Cedro, P. O. Box 6500, Carlsbad, CA 92008

NO POSTAGE
NECESSARY
IF MAILED
IN THE
UNITED STATES

BUSINESS REPLY MAIL
FIRST CLASS PERMIT NO. 271 CARLSBAD, CA

POSTAGE WILL BE PAID BY ADDRESSEE

Craftsman Book Company
6058 Corte Del Cedro
P. O. Box 6500
Carlsbad, CA 92008—0992

NO POSTAGE
NECESSARY
IF MAILED
IN THE
UNITED STATES

BUSINESS REPLY MAIL
FIRST CLASS PERMIT NO. 271 CARLSBAD, CA

POSTAGE WILL BE PAID BY ADDRESSEE

Craftsman Book Company
6058 Corte Del Cedro
P. O. Box 6500
Carlsbad, CA 92008—0992

NO POSTAGE
NECESSARY
IF MAILED
IN THE
UNITED STATES

BUSINESS REPLY MAIL
FIRST CLASS PERMIT NO. 271 CARLSBAD, CA

POSTAGE WILL BE PAID BY ADDRESSEE

Craftsman Book Company
6058 Corte Del Cedro
P. O. Box 6500
Carlsbad, CA 92008—0992